CW00402570

RIVENHALL

RIVENHALL

THE HISTORY OF AN ESSEX AIRFIELD

BRUCE STAIT

Edited by ANDREW STAIT

AMBERLEY

*To all the British, Commonwealth
and American men who flew from Rivenhall
and died in the cause for freedom.*

Bruce Stait 1928–1995

First published 1984
This edition 2011

Amberley Publishing
The Hill, Stroud
Gloucestershire, GL5 4EP

www.amberleybooks.com

Copyright © The Estate of Bruce Stait, 2011

The right of The Estate of Bruce Stait to be identified as the Author
of this work has been asserted in accordance with the
Copyrights, Designs and Patents Act 1988.

All rights reserved. No part of this book may be reprinted
or reproduced or utilised in any form or by any electronic,
mechanical or other means, now known or hereafter invented,
including photocopying and recording, or in any information
storage or retrieval system, without the permission in writing
from the Publishers.

British Library Cataloguing in Publication Data.
A catalogue record for this book is available from the British Library.

ISBN 978 1 4456 0403 9

Typesetting and Origination by Amberley Publishing.
Printed in Great Britain.

Contents

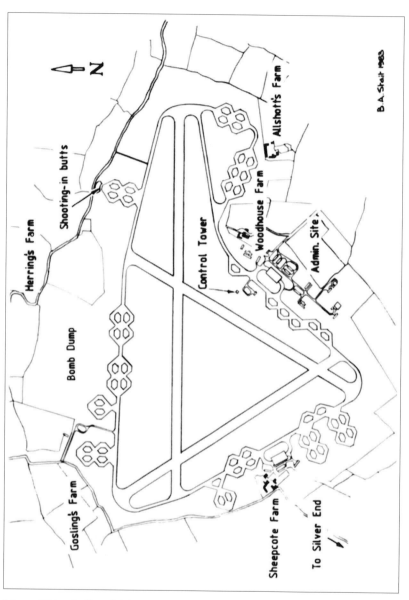

Map of Rivenhall airfield in 1944.

Foreword

The author confesses to a lifetime passion for anything connected with aviation, probably stemming from watching the 5/- (25 p) pleasure flights which operated from the beach at Southport in the summer holidays in the 30s. My family moved from Manchester to Essex in 1941 and as a schoolboy, in common with most other boys of my generation, I became intensely interested in all the aerial activity in the neighbourhood.

Regretfully, only an imperfect memory remains of those wartime years but with the help of many friends and colleagues and aided by the official records which are now available, a fairly accurate account of the squadrons and men who served at Rivenhall can be compiled.

I was born just too late to see wartime service, receiving my call-up papers in May 1946. When I joined the RAF, hundreds were in the process of being demobilised, among them a great many potential aircrew who never completed their flying training, proudly sporting a white flash in their caps. My 'demob' number was 75 (the highest is thought to be 81 – after that the Government introduced a statutory two-year service for all those eligible at the age of eighteen but minus the benefit of a free 'demob' suit at the end) and my service career ended in January 1949.

The research for this history began in 1975 with a letter to the local paper the *Braintree and Witham Times* and drew many responses from people with their own personal memories of Rivenhall airfield.

Where individual ranks are given they are as at the time of the events described. Many officers and men went on to achieve higher ranks in the post-war air forces.

Bruce Stait
Cheltenham

Introduction

Life in wartime England had few compensations, but to the many aircraft-mad schoolboys who lived through those momentous years, each day brought further joys to delight the eye. The skies were full of aeroplanes, particularly during the later years when the huge American aircraft industry got into its stride and began to pour out its products in an ever-increasing flood. The majority were destined for the new airfields in East Anglia and a number of them were within easy reach of the author's home village of Silver End. The nearest was less than a mile away and was always referred to by the local inhabitants as 'Silver End drome'. Indeed it came as something of a surprise when, several years later, the author discovered that it was officially named RAF Rivenhall due to its boundaries lying within Rivenhall parish.

Silver End could not be considered a typical Essex village as it possessed none of the thatched cottages and timber dwellings of nearby Cressing and Coggeshall. It was built in the 1920s as something of an experiment in self-sufficiency by the Crittall family, of metal window fame. Many of the houses have flat roofs, more suited to sunnier climes and the whole style of architecture now appears somewhat dated. It seems probable that few of the service personnel from the airfield were even aware of its existence. The GIs made the White Hart in nearby Braintree their meeting place when they could snatch a few hours away from camp. Longer leaves – or furloughs – were spent in London, or on occasions, as far away as Scotland. RAF types didn't seem to mind in which of the many pubs in the area they spent their off-duty moments.

Rivenhall rarely made the headlines, probably because it was one of the last British wartime airfields to be built. The publicity boys had concentrated their efforts on nearby Debden and Great Saling (Andrews Field) and few photographs or articles were ever published concerning the exploits of the Rivenhall squadrons. One exception occurred in July 1944 when staff from the magazine *Aeronautics* visited the base and flew with the B-26 crews. The photographs taken on that occasion by the celebrated Charles E. Brown are among the finest ever taken. He used some of the very early Kodachrome film and these superb colour shots are still being reproduced today by the aviation press.

Rivenhall airfield has no special claim to fame. Its operational life was less than two years, yet within that span a great deal was accomplished. Rivenhall squadrons took part in two of the largest air actions of the war; the invasion of Europe and the airborne crossing of the Rhine. It started life as an American fighter base, changing these for medium bombers after a few weeks and finally becoming the home for RAF secret operations squadrons which are still awaiting a full chronicle.

Of the four main aircraft types which flew from Rivenhall only the P-51 Mustang survived into the post-war era where it became front-line equipment for many of the smaller nations. In the 1980s, numbers of the bubble cockpit 'D' versions are still flown in the United States, in races and air shows but no flying examples are currently in Europe at the time of writing. Static exhibits are displayed in museums and the one owned by the Imperial War Museum at Duxford is a good example of an 8th USAAF Mustang.

From the total of 5,266 Marauders built, a few survive in the USA. The one owned by the Confederate Air Force is hoped to be restored to flying condition in the mid-1980s.

Sadly there are no Stirlings left, but in 1975 in a letter to *Flight International*, it was proposed to raise LJ899, a Stirling IV of

190 Squadron, from Lake Rydafors in Sweden in the same manner that the only surviving Halifax, now in the RAF Museum, was rescued by a team of divers from a Norwegian lake. It now seems that most of LJ899 was scrapped shortly after the crash and only a few parts were recovered in the mid-70s.

The Horsa gliders, being constructed mainly of wood and fabric, were unceremoniously burnt in their hundreds after the war ended, but parts of fuselages may still be found masquerading as garden sheds and chicken huts in rural areas. The Army Museum at Middle Wallop has a complete fuselage on exhibition.

Looking back is a pastime which most of us enjoy at some time or another and looking back to the war years is becoming increasingly popular. Schooldays have long had the reputation of being the happiest of our lives and certainly the author's schooldays were filled with many happy memories which have not diminished with the passing years. The train on which I travelled each day to Colchester was often the target for the local Mustang and Thunderbolt pilots who delighted in making low-level 'strafing' attacks similar to the real thing they were doing on the Nazi-run railways in occupied Europe. Even before leaving for school I could watch the silver Marauders as they took off in the morning sun, roaring low across the green fields of Essex as they set off for the bridges and the German flak defences that waited for them in France.

Official histories speak of the summer of 1944 as being one of the worst for weather in the twentieth century but schoolboy memories are of bright sunny days from dawn till dusk. We would swim in the nearby gravel pit or troop over to 'our drome' to watch the Americans and their wondrous flying machines. My brother, who was twelve at the time, struck up a friendship with the crew of one Marauder named 'By Golly' and used to come home with gifts of gum and bars of candy. A personal recollection of US hospitality is of attending a party given for the children of the district in the village hall at Silver End and of sampling the unfamiliar delights of American sweets and chocolate bars. To someone who had come to enjoy the rationed wartime Cadbury 'Blended' chocolate, the strange tastes were something of a surprise. The records of the 397th Bomb Group contain references to another party given for local orphans from Kelvedon in July 1944 but this took place on the airfield, almost certainly in the enlisted men's canteen.

Probably the most vivid memory for anyone who lived in wartime East Anglia was the flood of American servicemen who were to be found in towns and cities on every street, pub and café. Braintree was a popular centre for the 'Yanks' off duty and it was ringed with half a dozen nearby airbases, each of which disgorged hundreds of red-blooded Americans in search of relaxation and – if possible – female companionship. With their neatly tailored uniforms and easy-going manner and attitude they won the heart of many a young lady throughout the British Isles. I can remember a feeling of incredulity upon hearing that my mother had been 'chatted up' by some GIs during a cycle ride into Braintree – was nothing sacred! She was then at the ripe old age of forty-five.

One of the consuming passions for the youngsters of that period was collecting war souvenirs and the lengths to which they went in order to fulfil that passion still brings me out in a cold sweat whenever the subject is brought up. The brass cartridge cases from the .50 ammunition used by the American forces in their machine guns were a prized item, and the possession of a really long belt of these spent cases was the mark of an avid collector. I used to pull the bullets from live ammunition rounds with a pair of pliers, with a complete disregard for personal safety – or indeed that of others of my family. Just how these highly dangerous items came into my possession is not clear, but the district had its share of crashed aircraft which no doubt yielded such treasures to the youthful souvenir hunters. Children of the war years had only simple pleasures!

Building the Airfield

For over three years of the Second World War the Royal Air Force Bomber Command had been the only force capable of attacking the German war production industries. With the arrival of the USA in the struggle, the Allied policy of a strategic air offensive was reinforced but the build-up took time. American factories began turning out vast quantities of new and improved aircraft and a crash programme of airfield construction was undertaken in Britain to provide bases from which the new American Air Forces would operate.

The sites chosen lay mainly in East Anglia, in reasonably flat countryside with few large centres of population. As a bonus it also offered the shortest route for attacks on the German industrial centres situated in the Ruhr valley. The airfield building programme was the largest Civil Engineering project ever undertaken; the amount of concrete required was equivalent to 4,000 miles of three-lane motorway. Within the space of a few months, during 1942 and 1943, fifteen airfields were built in Essex, all of them in the northern half of the county. These were only a small part of the vast total of almost 500 airfields built in Great Britain during the war years.

Construction was under the direction of the Air Ministry Directorate General of Works, thankfully abbreviated to AMDGW but more familiarly known as 'Works and Bricks'. Many of the Essex airfields were built by US Army Engineer Battalions but Rivenhall was handled by a private contractor, W. C. French, and completed in 1943. Fields were levelled, drainage introduced, services installed, pipes laid and concrete poured. The average airfield required 130,000 tons of concrete and hardcore and 50 miles of pipe and conduit. The cost of constructing each flying field averaged £500,000 and the total cost, including the buildings and services, was in the region of £1 million (1943 prices).

On average each airfield took up approximately 500 acres and the loss of valuable food-producing land was extremely worrying to the authorities and even more so to the farmers who depended on the land for their livelihood. At Woodhouse Farm, on the very edge of the airfield, 270 acres were reduced to a mere 50. A fortunate local asset was the existence of large quantities of gravel, in easily accessible sites. Thousands of cubic yards were extracted from the Silver End gravel pit and, after the workings had been finished, it became a local swimming hole of great and occasionally fatal attraction, where the steeply sloping sides were a potential death trap for the young or inexperienced swimmer.

Of the hundreds of service men and women who were to pass through Rivenhall airfield during the following years, one man had the unique experience of being employed by W. C. French during its construction and two years later walked in through the main gate dressed in Air Force Blue. Mr F. V. Foreman, of Crich in Derbyshire, was involved in setting out the runways and perimeter track, 'before the proverbial sod was turned'. His expert knowledge of the layout of the airfield was to stand him in good stead in later years as we shall discover.

Rivenhall was laid out with the standard three runways of the period, one of 2,000 yards and two of 1,400 yards with a perimeter track and hardstanding for the aircraft. It was equipped with two T2 hangars, a multitude of Nissen huts for a variety of duties, sick quarters, mess halls, a cinema, firing range for testing the machine guns, armoury, workshops, stores, petrol storage, and a control tower. The bomb dump was situated on the north side, not far from the little hamlet of Bradwell and as far as possible from the living and working areas which were mainly on the south side of the airfield.

Arrival of the United States Army Air Force

The first unit to occupy the airfield was the USAAF 363rd Fighter Group of the US 9th Air Force commanded by Colonel John R. Ulricson. Trained in the United States on the P39 Airacobra, the group made the hazardous winter crossing of the North Atlantic on board the Queen Elizabeth and docked in the Firth of Clyde on 20 December 1943. Their first base was Station 471, Keevil in Wiltshire, where they arrived on 23 December 1943. A month later, on 22 January 1944, the group began the move to Station 168, at Rivenhall, and this was completed by 4 February. The group comprised three squadrons, Nos 380, 381 and 382 but initially they suffered from a shortage of aircraft. It was not until 24 January that they were allocated a total of eleven P-51B Mustangs and a training programme was quickly rushed into action.

Generally agreed to be one of the classic fighters of World War II, the Mustang was originally designed for supply to the RAF as the Mustang 1 (P-51A) powered with an Allison liquid-cooled engine. The initial testing and early service use proved disappointing but the decision to re-engine with a Packard-built Rolls-Royce Merlin resulted in a vastly improved performance and in this form the Mustang was supplied to every theatre of war as the P-51B and C (Mustang III) and P-51D and K (Mustang IV).

In the spring of 1944, the daylight bombing offensive, which had been instituted by the Americans in 1943, was resumed in earnest and the US 8th Air Force prepared to strike deep into Germany against the aircraft and armament factories. The difference now was that the bombers had adequate fighter cover all the way to the target, provided by the Mustangs fitted with drop tanks, and the Luftwaffe now found to its cost what it was like to lose one of the prime requisites of modern warfare – air superiority over one's own country. In order to inflict the maximum damage to the Luftwaffe, the Allied planners began to deliberately seek out targets and routes over Germany which would compel the defenders to do battle. The inevitable increase in American casualties in the constant attacks against Berlin in the first week of March 1944 were part of this costly strategy. The American losses were heavy. During the whole of the bombing campaign the US 8th Air Force lost a total of 10,000 men, 500 fighters and 800 bombers.

The losses for the US 9th Air Force were much lower; just over 3,400 were lost or killed in action. Aircraft losses were 2,944 – this figure included 2,139 fighters.

The Luftwaffe now began to experience the same problem which had faced the RAF in the summer of 1940, when a shortage of trained pilots became a vital factor in the Battle of Britain. Although fighter production actually increased during the last twelve months of the war – despite the Allied bombing campaign – the Germans were unable to train sufficient pilots to offset their losses at the hands of the long-range Mustangs, Thunderbolts and Lightnings. A further setback was the acute shortage of aviation fuel and oil which was a direct result of the Allied bombing programme.

The P-51Bs of the 363rd were initially painted olive drab with grey undersides with approximately 18 inches of the nose painted white. When replacements were received in natural finish, this area was painted black. After they had left Rivenhall each of the three squadrons had different colour noses: 380 (coded A9) had a blue band, 381 (coded B3) were yellow and 382 (coded C3) were red. So far as can be established the only replacement aircraft at Rivenhall were P-51Bs; the bubble cockpit version, the P-51D, did not start to arrive until some time in May, by which time the 363rd had left Rivenhall.

Twenty-seven pilots from the 365th FG were transferred to the 363rd on 5 February to bring all three squadrons up to operating strength and the group put in a great deal of training during the remainder of the month. The group records do not mention the concentrated efforts of the bombers of the US 8th Air Force during the period which became known as 'Big Week', from 19 to 25 February when a total of 3,300 bomber sorties were flown, escorted by the long-range fighters of the 9th AF. It was during this week on 22 February that they were scheduled to perform their first 'affiliation' mission in preparation for a planned combat mission the following day. Unfortunately the weather caused their first effort to be something of a fiasco. After an early setback a total of twenty-four Mustangs took off led by Capt. Culberson, the CO of the 381st Sqdn but a sudden snow storm caused an early return and all the pilots landed safely. The weather the next day had still not improved sufficiently and it was not until the 24th that the 363rd flew their first 'proper' mission.

The 363rd FG was the third Mustang group to be formed in England. The honour of being the first went to the 354th FG, who had pioneered Mustang operations throughout the winter months from their base at Boxted near Colchester and had already earned a high reputation for their aggressive actions against the Luftwaffe.

It was decided to use the experience of the 354th Group to give some tuition to the fledgling fighter pilots of Rivenhall and accordingly Major James H. Howard flew over from Boxted to lead the new group on a withdrawal support mission, rendezvousing with the bombers in the Brussels region. Major Howard was one of the most distinguished fighter pilots of the Second World War. He was the CO of the 356th Fighter Squadron and had been awarded the Congressional Medal of Honor, the highest American award for bravery for single-handedly defending a B-17 formation from enemy fighter attacks on 11 January 1944 and in the process shooting down three and damaging several more.

On 24 February mission two FW190s were sighted but no attack was made and the only incident was to Lt John W. Schmidt's Mustang which received slight flak damage. The 363rd flew two more uneventful missions on 25 February, both of them led by another of the pilots from Boxted, Capt. Jack T. Bradley, who was credited with five and a half victories at the time and went on to increase his score to fifteen by the end of hostilities. The last mission of the month took place on 29 February led once again by Major Howard but not without incident. Two of the Mustangs collided during the rendezvous and although extensively damaged, both pilots returned home safely.

1 March 1944 was the first anniversary of the formation of the 363rd FG and the following day, with Capt. Bradley in the lead, they flew their next mission as withdrawal support for the 8th AF bombers. There was some very accurate flak in the notorious Ruhr Valley which holed a few aircraft and Lt Tyler of the 380th Sqdn ran out of fuel on his way back and belly-landed at Boxted, fortunately without injury.

It was on Friday 3 March that the group were finally tested in combat, when they acted as part of the escort for the first American daylight attack on Berlin. The capital of the Third Reich had received many night-time raids from RAF Bomber Command (at one time the city was raided on 200 consecutive nights) but the poor weather in the winter of 1943/44 had prevented any daylight attempts, plus the lack of a long-range fighter escort. A total of thirty-six Rivenhall Mustangs took part but eleven had to turn back for a variety of reasons. They were being led for the last time by Howard, now a Lt-Col., and it was largely thanks to him that the group achieved a classic 'bounce' when they encountered a mixed force of some thirty-plus Messerschmitt Bf109, Me410 and FW190 fighters about 3,000 feet below them. In the heat of the moment several pilots forgot all they had learned in months of training; Major Culberson released his drop tanks and forgot to switch over to the main supply with the result that his engine almost cut out, and Lt Boland lowered his wheels instead of dropping his tanks. Nevertheless, the group claimed the destruction of a Bf109 (Lt Brink) and a Me410 (Major Culberson) with three Bf109s damaged or probably destroyed.

Berlin had been lucky once more – the bombers were recalled due to the mountainous cloud over Germany but the fighters were already at the rendezvous point when the recall signal was made and the German fighters appeared shortly afterwards.

In complete contrast to the success experienced by the jubilant pilots, the following day was to go down in the annals of the 363rd as the blackest they ever had. Thirty-three Mustangs took off on 4 March to give penetration support for the heavies of the 8th AF, who were making a further attempt to hit 'Big B'. A thick layer of cloud covered most of Western Europe and once again the recall order went out. Some bombers never received the order and they went on, with the Mustangs attempting to perform their escort duties in spite of the worsening conditions and these few B-17s eventually dropped their bombs on Berlin. It is probable that the German tracking service made skilful use of the towering clouds to vector the Bf109s of the elite 11/JG.1 into the area around Hamburg where they ambushed the scattered fighter escort with terrible results for the pilots of the 363rd. Of the thirty-three Mustangs despatched, eleven aborted for a variety of reasons – the Mustang had more than its fair share of mechanical troubles in the spring of 1944 – and only eleven made it back to Rivenhall. The loss was almost equally divided, five from 381 Sqdn and six from the 382nd. There was much speculation about the missing men, but none of the returning pilots could offer an explanation for the disappearance of a third of the group. On the next day there was no-one above the rank of First Lt to lead the group; those eligible were reported sick and in hospital.

The mystery of the missing Mustangs continues to be a source of argument and conjecture among historians. Opinions are fairly evenly divided between the 'ambush' theory and the other possibility that they became disorientated in the thick clouds, ran out of fuel and ditched in the North Sea. This latter view is supported most strongly by the survivors who came back from the 4 March mission.

Nevertheless, the 363rd performed their duties on the mission of 5 March, acting as escort for a force of bombers flying 400 miles to Bordeaux. It was a completely uneventful mission except for

Lt Vance of the 380th Sqdn. He was almost to the rendezvous point when his engine began to run rough, but by careful management he nursed his aircraft back to Rivenhall, where the cause was quickly diagnosed – faulty spark plugs. This was to be a common source of complaint and was almost always cured by replacing the existing plugs with a suitable British alternative. That evening there was a party to celebrate recent promotions among the officers.

Because of their tragic losses on 4 March the 363rd missed the costliest attack ever flown by the US Army Air Force on 6 March when a total of eighty heavy bombers were lost and many more damaged, during a further attempt to bomb the German capital. Despite a large escort of fighters of the US 9th AF, the German fighter pilots succeeded in penetrating the screen and gave the Americans a severe defeat, but the tide was now beginning to turn against the Luftwaffe.

The 363rd returned to their escort duties on 8 March, when a force of 600-plus B-17s and B-24s mounted yet another attack against Berlin. Lt O'Connors of the Boxted Group led them in a clash with a mixed force of Bf109s and FW190s which they encountered near the target and claims of five enemy fighters destroyed were made at the debriefing. Second Lt Neill Ullo from the 380th Sqdn was their only loss; he was last seen with his tail shot away and his plane disintegrating around him. A total of thirty-seven bombers were lost, considered to be an acceptable loss rate.

On the next day, Berlin was once again covered by 10/10 cloud and the Luftwaffe wisely stayed on the ground. The flak gunners downed nine of the heavies but the 363rd fighters returned to Rivenhall unscathed. Bad weather was the cause of many of the large-scale daylight attacks being aborted during this period and the 363rd spent a great deal of time on 'Maintenance and Training', an official term for hanging about, waiting for an improvement in the weather. On nineteen days in March the weather was too poor for the group to undertake combat missions and even those which did take place were often thwarted due to unforeseen weather fronts moving in.

For eight consecutive days, from 10 to 17 March, the pilots of the 363rd chafed at their enforced idleness and looked at the sky. One bright spot was the arrival of Brig. Gen. Quesada on 13 March, the Commander of the US 9th Fighter Group piloting his own natural finish P-38 Lightning. Quesada distributed eleven decorations to twelve pilots at a ceremony held in spring sunshine in front of the hangar. The group official history records that some fancy footwork and fast talking took care of the odd man out. The month of April started with an escort mission to Ludwigshafen but clouds up to 21,000 feet over Germany caused the recall signal to be sent out. The 363rd were escorting a group of B-24s which continued heading into enemy territory, where they became hopelessly lost, eventually bombing Schaffhausen in neutral Switzerland. The six-hour mission was described as a 'Cook's tour of Europe' by the returning pilots.

Poor weather continued to restrict operations for the next three days, then on 5 April the group went out on a strafing mission against enemy airfields in France but cloud at the target area caused them to look for alternatives. They found Triqueville but the defending flak hit the Mustang of Lt Lewis of the 382nd, and he bailed out.

Good Friday, 7 April, was a very different day for the 363rd; they were treated to an air display by three German aircraft, a Junkers Ju88, a Messerschmitt Bf109 and a Focke Wulf FW190, all part of the Enemy Aircraft Flight which flew from RAF Collyweston. The author recalls the occasion vividly; the day was dull and overcast with a light drizzle, much like public holidays of the present day! The 363rd put up top cover to prevent any accidents and thoroughly enjoyed the spirited performance of the RAF pilots in the German machines. The Enemy Aircraft Flight made regular tours of bases in the UK, displaying captured examples of German aircraft painted in RAF markings and camouflage colours, yellow undersides and dark green and dark earth on top. On this tour they took in the airfields at Great Dunmow, Andrews Field and Earls Colne as well as Rivenhall.

In their final week at Rivenhall the group suddenly became engaged in a flurry of activity. On 8 April they escorted part of a

600-strong formation of B-24s to Brunswick where they fought off a strong attack by Bf109s and FW190s in large numbers. Seven were claimed destroyed and six damaged but the group lost Second Lts Fontes and Wenner and six B-24s were also lost. The following day the heavies went on one of the longest flights to date, against Pozan and Marienburg and achieved a good bomb strike. Two more pilots were lost from the 363rd, Second Lts Steinke and Pollard, both going down in the North Sea for no obvious reason.

Two missions were flown on the 10th; in the morning the group dive-bombed the marshalling yards at Hasselt in Belgium, where Col. Ulricson scored a direct hit on the engine and front cars of a train. In the afternoon they escorted a Marauder group of the 9th Air Force against another marshalling yard, this time in France.

Another long escort mission with the B-17s to Berlin provided the 363rd pilots with an opportunity to indulge in airfield strafing. Felix Kozaczka flying with the 382nd describes the events of 11 April:

On the way home we went back down to the deck and looked for likely strafing targets. We came across the airfield at (I believe) Burg, near the Elbe, and went across line abreast in flights of four. I picked out a He177 as my target and blasted away. After the mission, the interrogator asked me how many more aircraft I saw besides the one I was strafing. I believe I told him that I only saw the one that my element leader was shooting at and one or two more. However, when my gun camera film was developed, we could count 16 fighter aircraft between the airfield boundary and the He177 I was shooting at!

Second Lt Boatright wasn't so lucky, a 20-mm shell exploded in the cockpit, severely wounding him in one eye. Despite the injury he regained control and held formation until the squadron got back to England.

A mix-up on 12 April resulted in the loss of another pilot from the 381st. The bombers were recalled because of bad weather but the escort flew on to the rendezvous point and beyond. In the vicinity of Magdeburg, Second Lt Howell of the 381st disappeared. Strangely, he was the co-owner of a monkey with Lt Steinke, who had been lost on 9 April. Their final mission from Rivenhall was flown the next day when they escorted the heavies to aircraft plants in Southern Germany. Claims were made for two Bf109s by Lts McGee and Schmidt. Some hectic packing followed their return and the group moved to Staplehurst, an advanced landing ground in Kent on Friday 14 April.

In order to provide fighter cover for the Allied invasion fleet which was assembling in ports along the south coast, the 9th Air Force transferred all its fighter units to airfields in that vicinity during the month of April. On 18 April three of the 363rd Mustangs landed at Rivenhall, for old time's sake, following an escort mission.

The 363rd Fighter Group went on to finish the war in Europe, operating from airfields on the Continent. From 4 September 1944, they became a tactical reconnaissance group equipped with camera-carrying F6C Mustangs and many of the original pilots took transfers to other fighter groups in order to stay with the action.

The Bridge Busters

Saturday 15 April 1944 saw the arrival at Rivenhall of the 397th Bomb Group (Medium) under the command of a thirty-three-year-old Texan, Colonel Richard T. Coiner Jr. The four squadrons which formed the group were among the last to be activated by the US 9th Air Force and many of the personnel, who had instructed on the B-26 training programme in the United States, were beginning to worry that they might miss taking a more active part in the war.

The group was equipped with the Martin B-26 Marauder which had originated from a US Army Air Corps specification for a fast five-seat medium bomber and had first flown on 25 November 1940 (the same day as another famous twin-engined machine, the prototype Mosquito). The Marauder was powered with two Pratt and Whitney eighteen-cylinder air-cooled radial engines and it was armed with eleven .50 machine guns. Its maximum speed (at maximum weight) was 274 mph at 15,000 feet with an endurance of just over five hours. These were the figures for the B-26G, the variant which was eventually to equip most of the groups but the 397th came to England with the B-26C version.

Some of the officers had plenty of wartime or civil flying experience behind them. Lt-Col. Franklin Allen, the CO of the 598th BS had had a distinguished career flying the B-26 in the Pacific while Major 'Casey' (K. C.) Dempster, the original CO of the 597th BS was a pre-war veteran and had taken part in the campaign in the Aleutians.

He was succeeded in January 1944 by Lt-Co. Frank Wood, who had flown as a co-pilot with Pan American Airways. The crews flew their aircraft to nearby Gosfield (Station 154) via the Southern route (South America, Ascension Island, North Africa) and began to arrive on 8 March. The ground crews departed New York on the ex-Italian liner *Saturnia* on 23 March and eventually joined up with the aircrews on 5 April 1944. During this lengthy wait for the ground echelon to arrive the group flew practice missions whenever possible. Each squadron numbered some sixty to seventy officers and around 300 enlisted men. Combat crews were initially a total of fifteen crews for each squadron but were almost immediately increased to twenty-one, followed by a further increase to a total of twenty-four.

The Marauder had become a common sight in the district long before the 397th came on the scene, as one of the first groups to equip with this type had been the 322nd at Great Saling, known to the Americans as Andrews Field. To them had befallen the task of pioneering tactical daylight bombing over occupied Europe at medium altitude, for which the Marauder, after some initial setbacks, had proved eminently suitable. There had been a number of criticisms of the aircraft in the beginning and indeed several attempts had been made to cancel the programme in favour of more docile types. Fortunately, despite the critics, the B-26 was finally built in large numbers and went on to become one of the great aircraft of World War II.

The 397th Marauders were initially painted olive drab on upper surfaces with neutral grey undersides, and a yellow diagonal band about 21 inches wide painted across the fin and rudder, which also carried the serial number in black. Original serial numbers were in the range 42-96029 to 125, the last three digits being the common form of reference. From 42-96129 onwards the aircraft dispensed with the camouflage finish and were left in natural metal finish. The popular press dubbed them 'The Silver Streaks'.

The squadron codes were as follows: X2 – 596, 9F – 597, U2 – 598, 6B – 599. (Some reference works have quoted incorrect codes for the 397th but those given here are from the official records.)

25

The group settled in quickly and on 20 April 1944, just five days after their arrival, they took off on their first mission. Led by Col. Coiner, with 'Casey' Dempster leading the second box, the thirty-six aircraft set course for the target, a launching site for the V1 flying bomb, located at Le Plouy Ferme on the Pas de Calais. At this time the Germans were still perfecting the weapons and no V1s had been launched against England, but the existence of Hitler's secret revenge weapon had been known to Allied Intelligence for some months. Close scrutiny of aerial photographs taken over Northern France had revealed an increase in building activity over a wide area, all having common features, the most ominous being an inclined ramp pointing at London. Alongside was a curious curved building, resembling a ski laid on its side (it was used for storing the V1 prior to firing) which prompted the use of the term 'ski site' for all these locations.

The fascinating story of 'Crossbow', the code name for Allied operations against the secret weapons (including the V2 rocket) has been fully documented in many accounts but as these were the first targets for the 397th it may be useful to give a short account here.

Brave acts of courage by the resistance movement who provided information and parts of the V1, plus inspired interpretation of photo reconnaissance prints, helped to put together a picture of what might lie in store for the people of England should the German weapon programme succeed. A massive bombing campaign was begun by the Allied Air Force on 5 December 1943 when a total of sixty-nine ski sites had been identified. By the end of May 1944 the Royal Air Force and the US 8th and 9th AF had plastered the sites (totalling 140 by that time) with thousands of tons of bombs. The results, according to official estimates, were that 103 were unfit for use and most of the others had been severely damaged. Unfortunately this optimistic picture was only part of the full story.

Since the original sites were obviously vulnerable to Allied bombing the Germans had developed alternative launching arrangements using a prefabricated and less permanent construction. The ski sites were abandoned at the beginning of January 1944 but they continued to receive attention from the Allied bombers, while the modified sites progressed unscathed and unseen. In fact it was not until May that modified sites began to be identified from the photo covers, but due to extreme difficulties very few attacks were successful or indeed attempted. Pilots and bombardiers had to correctly pick out the well-camouflaged sites which were frequently merged into the French farms which abounded in the area, and the task became increasingly difficult. The V1s were supplied from storage sites, some of which had been identified and these also came in for extensive bombing, but once again the Germans countered the new threat by moving the supply sites underground, unbeknown to the Allies who continued to attack the original locations. That at Beauvoir received a hammering from the 397th and other 9th AF groups on 29 May. In the first two weeks of June a crescendo of bombs rained down on the supply and launching sites in what was now a largely valueless exercise.

It was not therefore surprising that despite the weight of the combined Allied bombing campaign the first V1 flying bomb fell on the British Isles in the early hours of 13 June and from then on the firing rate increased steadily. Within a fortnight 2,000 V1s left their firing ramps and an average of 120–190 were fired per day between June and September, not all of which reached the British Isles.

Obituary notices which might be helpful to the enemy, were forbidden in all local and national newspapers; instead they referred in vague terms to 'Rocket attacks by flying bombs in Southern England'. After the war the Fire Forces Commander P. G. Garon MC, GM commented wryly, 'Essex took up a devil of a lot of Southern England during the V1 attacks.'

In fact Essex received a total of 412, the majority probably air-launched from converted Heinkel bombers from the comparative safety of the North Sea. The Greater London area had by far the worst of the V1 attacks but the adjoining counties all suffered from the strays or under or overshots. Surrey had 295, Sussex 880 and Kent headed the list with 1,444.

Although the 'doodlebug' – the name which was coined by the civilian population – caused enormous damage to property, the loss of life was relatively small. On average each bomb killed only two people and injured five, but there is little doubt that had it not been for the efforts of the defenders in repulsing so many bombs before they reached their target, the total effect would have been substantially worse.

The 397th attacks on 'No-ball' targets (the Allied code name for the launch sites) at the end of April were frequently disappointing, being classed as either 'poor' or 'good' as the crews accustomed themselves to playing their part in the European Theatre of Operations (usually abbreviated to ETO).

A stroke of good fortune occurred on their fourth mission when attacking a coastal battery at Benerville. The strike photos showed one of the bombs had landed some 700 yards offshore, producing a series of fourteen small explosions which revealed the existence of underwater obstacles in the shallows, thus confirming Allied suspicions regarding the German defences on the Normandy beaches.

It was not until mission No.7 that the group had its first notable success when they attacked beach defences near Oustrieham on the Normandy Coast, shortly to become famous as SWORD beach at the extreme eastern end of the invasion area. Ninety-seven per cent of their bombs were judged to have landed in the target area.

The bombing campaign in Normandy in support of the forthcoming invasion was now of prime importance so that gradually the attacks on the V1 sites were eased off. In their place the fighters and AA batteries took over and began to get the upper hand. Of the 7,547 ramp-launched V1s, a total of 3,957 were destroyed, over 3,600 by fighters and AA fire, almost equally divided. An unknown number were also air launched, possibly over 1,000.

A vitally important part played by the Allied Air Forces in the weeks prior to the invasion of Europe was the 'interdiction programme'. Carefully planned attacks were made against vital road and rail bridges and centres of transportation, to seal off the proposed combat area while the Allied armies built up their strength on the enemy shore. In order to keep the Germans in the dark as to

Allied strategy, twice as many bombs fell on targets in the Pas de Calais area. All the bomber groups in the US 9th AF were called on to play their part in this campaign and the 397th was no exception. More than a third of their missions during this period were bridges and their unofficial title became 'The Bridge Busters'. The bridges were frequently heavily defended. On their very first 'bridge busting' mission on 8 May, the 397th suffered its first loss over enemy-held territory when aircraft 143 piloted by Second Lt Freeman of the 596th peeled out of formation with both engines smoking after being hit by flak. Escorting Spitfires reported six parachutes, the whole crew, who all subsequently became POWs. Altogether twenty-six planes were damaged during this attack, the fourteenth mission of the group.

The next day saw the first fatality for the 397th when the group were assigned a No-ball target at Le Grismont in the Pas de Calais. Lt Thompson's aircraft was hit by flak and Second Lt Frank Evanick, the bombardier, was killed when a fragment penetrated his compartment. With the co-pilot wounded and severe damage to the hydraulics and rudder and elevators, the plane returned to Rivenhall and landed safely. 'Tommy' Thompson had a bad spell in May when on four occasions his aircraft received widespread damage, but each time his skilful airmanship got the crew home. During the mission of 24 May against the well-defended harbour at Dieppe, his aircraft was hit in the main fuel tank and other parts, but he pressed on with the attack. His right tyre had also been punctured, which caused some anxious moments when landing at Rivenhall. Those who took part in the Dieppe mission spoke of it as one of the hottest places they'd visited so far. The 397th rode through an intense flak barrage and many were hit. One of them was No. 166 flown by Lt Gross (who will figure again later in the narrative). His aircraft was badly damaged on the bombing run and the co-pilot, Lt Chadbourne, was seriously injured. They staggered back to England and Gross crash -landed at Newchurch.

Thompson was in trouble again two days later on 26 May when attacking the Luftwaffe airfield at Chartres. One of his engines began running rough soon after turning away from the target and he had to feather the propeller. Three of the Mustang escort kept him company on the two-hour flight back, on a single engine, and he landed safely at Rivenhall.

Bridges at Lieges, Le Manoir, Orival and Maissons Laffitte were attacked between 25 and 29 May, fortunately without loss, but many aircraft were damaged and required extensive repairs. One of the most important targets selected by the 9th AF planners was marshalling yards at Mantes-Gassicourt, on the River Seine near Paris, and the 397th had three attempts at removing it from the list. The first two on 28 and 29 April both proved abortive due to the heavy clouds which covered the area and the group returned without dropping their bomb loads. On 1 May a total of 149 1,000-lb bombs landed on the target and all aircraft returned safely. It was not always so. Another bridge near Paris at Maisons Laffitte was the target for the mission on 28 May when the crews reported the thickest flak to date and, although suffering no casualties, twenty-one aircraft were damaged. On a return visit on 24 June the group were not so lucky. The German flak defences were certainly up to the mark that day. They knocked down four aircraft from the first box but this time missed the lead aircraft. Three of those lost were from the 597th BS (Nos 120, 121, 127) flown by Capt. Gatewood, Lt Knox and Lt Neill and the fourth was No. 177 piloted by Capt. Kenneth Powers, of the 599th BS. The co-pilot of a lead 'window' ship, F/O Eyges, was killed and two aircraft (Nos 133 and 161) crash landed in friendly territory. Of the thirty-nine aircraft which took part in this costly mission a total of thirty-three were damaged and twenty-five crew members (of the aircraft which returned) were wounded.

One of the aircraft on this mission (possibly 'Mama Liz') was piloted by Capt. Moses Gatewood, who was flying as deputy leader in the last flight over the target. This had given the flak defences time to get the range and bearing and a thick and accurate barrage was being flung up. In the run up to the bridge, Gatewood's aircraft had already received minor damage, but as they started on the bombing run the ship was hit by a close burst, shattering part of the perspex

nose and the cockpit windscreen. Despite this setback Gatewood
continued his run and dropped on target. They were flying at about
10,000 feet and as they turned away another hit cut the elevator,
rudder and aileron cables and the aircraft dived out of control.
Gatewood rang the alarm bell and ordered the crew to bail out,
but the violent manoeuvres the ship was making prevented them
from complying with the order. Fighting hard to regain control,
Gatewood, by judicial use of the elevator trim tab, managed to
regain an even keel and very gingerly he began to steer away
from the target and head for the Normandy beachhead which lay
180 miles away. They were down to 4,000 feet and just when the
crew began to think they had a chance, another flak burst hit the
starboard engine, followed by another which severely damaged the
fuel tanks. Gatewood realised they were now in desperate trouble.
As he was unable to feather the propeller of the inoperative engine,
the aircraft quickly began to lose altitude. As more flak bounced
them around, the smoke and fumes started to make conditions inside
extremely hazardous. Gatewood gave the second order to abandon
ship. They were now down to 2,500 feet and losing 800 feet per
minute. The three crew members in the rear, including the wounded
radio operator who had been hit in the legs, successfully bailed out
leaving Gatewood, his wounded co-pilot and the bombardier. Unable
to open the bomb bay doors (the normal escape route) they tried to
open the nose wheel door. The bombardier finally achieved this by
jumping on the wheel and he and the co-pilot left through the door.
Gatewood removed his flak suit, helmet and headphones and quickly
followed them. By this time they were very low, probably under
1,000 feet and he pulled the ripcord at once. The parachute just had
time to open before he hit the ground and he rolled along between
two rows of apple trees. He had landed somewhere in the vicinity of
Rouen and he was a little surprised to find himself in one piece after
the hectic activity of the past few minutes. Disengaging his harness
and rolling up the silk, he ran towards a French woman who stood
with her bicycle on a nearby road, watching the proceedings with
interest. The woman burst into tears when, in schoolboy French,

he asked her for help, and an agitated farmer came running down the hill with wild gesticulations and shouting, 'Les Allemands, les Allemands!' Gatewood fled from the scene and hid in the nearby woods while the Germans scoured the area without success.

The remainder of his exploits in Nazi-occupied France read more like a Hollywood movie script and will only briefly be retold here. He wandered around in the woods for several days, seeking help but being refused on many occasions before coming across a Frenchman who was made of sterner stuff. Contact was made with the French underground resistance movement and he was whisked off to Paris, where he stayed for two weeks. In company with two men and a woman he left there by car on 13 July and headed for Marseilles. At every road block Gatewood expected to be arrested but their fake papers got them safely through. In the lovely old town of Avignon they were mixed up in a fight between the Maquis and the German occupying troops which finally culminated in them being hauled before the SS town commander. Incredibly, Gatewood and his companions talked their way out of that tight corner and after several further adventures he crossed the border into Spain. On 20 August he arrived back in England and eventually, a year later, he was awarded the Silver Star for his achievement.

Field Marshall Rommel, in charge of the German forces in the Normandy area, had his HQ on the banks of the Seine at La Roche Guyon and frequently witnessed the efforts of the American bombers to destroy his vital bridges. The last Seine bridge to go was that at Mantes Gassicourt on 30 May. The interdiction campaign was so successful that in the three months before D-Day only four of the eighty special rail targets escaped serious damage and the Germans finally abandoned all attempts to keep account of the damage and destruction. Rail traffic over the whole of France declined by 70 per cent.

Despite the fierce flak often encountered, the 9th AF Marauders had an exceptionally low loss rate. A total of more than 250 are reputed to have reached the 100-mission mark in the ETO. When considering this figure it should be remembered that the 9th were not involved in the deep-penetration raids of their 8th Air Force comrades, where the attaining of 100 missions was something of a miracle.

In common with other B-26 groups the 397th flew the standard two- and occasionally three-box formation of thirty-six aircraft, with the second box flying slightly below and behind the first. The Marauder was capable of carrying a 4,000-lb bomb load made up of either 250-, 500-, 1,000- or 2,000-lb bombs depending on the target and weather conditions. Bombing was generally a height of between 8,000 and 12,000 feet, again depending on conditions.

Aircraft flying in the lead position carried extra equipment and crew members specially trained in navigation in order to locate the correct target. These aircraft were often the ones which suffered most from the attentions of the German flak. 13 May was certainly an unlucky day for one such crew when Lt-Col. Frank Wood, the CO of the 597th BS, led the second box in an attack on coastal defences at Gravelines, near Dunkirk. His plane was extensively damaged by very accurate flak but Wood fought to maintain control and brought it back to make a crash-landing at West Malling in Kent. Three crewmen were wounded and Lt Evans was killed in this action. It was the first squadron fatality. Few lead crews returned without damage and many were shot down.

Operation OVERLORD, the Allied invasion of Europe, took place on Tuesday 6 June 1944 after a delay of twenty-four hours occasioned by uncertain weather conditions in the English Channel. In order to preserve secrecy, extra precautions were taken on every airfield in Britain on 5 June and among the hundreds of civilians who were detained until the last plane landed on that Monday evening was a lady who found the experience quite enjoyable. When she was questioned by a reporter from the *Essex Chronicle* on her enforced stay on an unnamed American base, she was full of glee: 'I had chicken for dinner!' Wartime rationing was always adequate but there was a marked contrast between the food enjoyed by American servicemen and the average British family. The many local people who befriended the GIs in the vicinity of the base were pleasantly surprised with gifts of canned food etc. which the British housewife could only obtain on rare occasions.

To ease the problems of aircraft identification for the Allied gunners manning the defences of the invasion fleet, all aircraft taking part were painted with alternate black and white stripes on the wings and fuselages. The 'invasion stripes' were applied on 4 and 5 June in a mammoth painting exercise which occupied most of the two days and continued into the evening prior to D-Day. It was the first clue to the nearness of the invasion for many of the crews who were to take part. The 397th were neatly painted but photos of other Allied aircraft sometimes reveal a lack of time or skill on the part of the ground crew who performed this task.

The group put up a maximum effort for the first of their two missions on D-Day, when fifty-three aircraft in three boxes, led by Lt-Cols McLeod, Berkenkamp and Allen, took off at 4.45 a.m. to join the thousands of Allied planes making their way across the English Channel. The targets for the 9th AF medium bombers were three coastal batteries and seven defended positions covering a beach on the Normandy coast that no one had heard of until D-Day. From this day onwards it became world famous under the code name of UTAH beach.

Jim Russell, the waist gunner in 'Dottie Dee', named after the wife of the pilot, recalls the mission.

About 10.00 p.m. on 5 June a major stuck his head in the Quonset hut that I and a couple of dozen other crew men knew as 'home' and said, 'Better hit the sack early you guys, I've a hunch there's going to be an early briefing tomorrow.' It was the earliest briefing I ever had. We were roused at 2.00 a.m. and after a hurried breakfast of powdered egg and toast we gathered in the briefing room in front of some top brass. Col Coiner was standing alongside a map of Western Europe and introduced General Sam Anderson, the 9th AF Bomber Command leader, to the assembled company. Coiner – a man with a flair for the dramatic – went on: 'Gentlemen, this is it – we are going to spearhead the invasion.' I had no wish to argue with such distinguished company but looking out at the dreadful weather conditions it certainly seemed unlikely that there would be an invasion or that we would spearhead it. But the briefing officer made it plain that this was a day like no other and that regardless of the elements Allied planes would fill the sky that morning and fill the sky they did. We took off before dawn, finally pierced the solid overcast and looked round for the rest of our six-plane flight. (In spite of the 24-hour postponement the weather conditions were still far from ideal and heavy clouds covered the whole of the invasion area.) As they seemed to have lost themselves we attached ourselves to another group of B-26s. It may have been an unorthodox procedure but it was something of an unorthodox day. Midway across the Channel we headed down through the overcast; it was daylight now and as I looked through the waist gunner's window I could see that there below was a vast armada of ships of all shapes and sizes, all sailing steadily towards the French coast. Roelif Loveland, a war correspondent who was riding with us wrote: 'We saw the curtain go up this morning on the greatest drama in the history of the world.' The navy began shelling the distant shore where the waiting German troops manned the coastal guns and beach fortifications and we could see the the explosions as the shells struck home. Because of the weather we were flying at 3,500 feet, far lower than normal, and the German troops opened up with machine guns but to no effect. In fact the ground fire was far less than we were accustomed to on our previous missions and the Luftwaffe did not put in an appearance at all.

For a few minutes the airspace over UTAH beach became the most crowded in the world as 276 medium bombers converged on the German defences. Each of the Marauders carried sixteen 250-lb bombs and at 6.21 a.m. they released on target. The timing was crucial. At 6.30 a.m. the first of the American troops waded ashore on to UTAH beach. Thanks to the bombardment by ships and planes, of the thousands who landed there that morning, only twelve were killed and just over 100 wounded, a remarkable achievement.

Flak damaged six aircraft of the 397th Group but it did not deter the thirty-eight aircraft which took part in the second D-Day mission, an attack against coastal defences at Trouville at the other end of the invasion beachhead.

An unusual incident occurred the following day on 7 June. Shortly after taking off to bomb the marshalling yards at Fiers the pilot of a 599th BS aircraft, Lt Kretschmer, felt a stream of tracer hit their ship, not from a German fighter but from the guns of the aircraft flying on their beam. A runaway turret gun had malfunctioned and accidentally sprayed them with a lethal burst of .5 shells at very close range, causing a major fire to break out. Kretschmer jettisoned his load of sixteen 250-lb bombs in open country and then performed the difficult feat of setting the blazing B-26 down in farmland not far from the English town of Rye in Sussex. All the crew were safe, with only minor injuries.

Lt M. Forey of the 597th BS was killed the next day, 8 June, in a crash at Chipping Ongar. The pilot, 'Flat Top' Cordell and F/O Breen (Navigator?) were also injured. The cause of the crash is unknown. They were not on the mission list for that day and it is possible that it was a training accident.

On 17 June Capt. Gus Williams was the only survivor from a ditching in the Channel following an attack against a bridge at Chartres. He was in the water for over three hours but, being a strong swimmer, he made it to the English shore.

Pathfinder missions had begun on 22 May with an attack on St Marie au Bosc with indifferent results, but from 22 June they began to take place more frequently. With this technique the responsibility for accuracy rested with the bombardier in the lead aircraft. When he released his bombs the following aircraft also toggled theirs and the resulting bombing pattern was found to be of a high standard of accuracy.

In the two months following D-Day another forty-five missions were flown by the Rivenhall Marauders, over half of them against bridges. Up until this time, most of the bridges had been along the River Seine, now the 9th AF turned their attentions to the River Loire which lay to the south and one by one the road and rail bridges disintegrated in a welter of accurately placed bombs.

On Sunday 16 July the group set off for another bridge at Nantes on the River Loire. The lead aircraft in the second box, carrying a nine-man crew, was piloted by Capt. John Quinn West. Capt. West was a deeply religious man, much respected by his colleagues and in particular by his crew. He was, by any standards, a severe man with a faith in God which precluded any forms of human weakness or frivolity. He was never known to drink or smoke and even dancing was on his list of prohibitions. Despite these things, or perhaps because of them, he was admired by all who knew him. It was his practice to offer a prayer for their safe return before every mission. His strongest epithet and the name which his crew chose for their aircraft was 'By Golly'.

The bridge was very heavily defended and the flak was some of the worst they had ever encountered; in no time they were in deep trouble with the right engine out and hydraulic lines spewing fluid out everywhere. In an attempt to lighten the load, the crew began to throw overboard as much equipment as possible. Gradually they regained control of the stricken aircraft but as they approached the Normandy coast it became obvious that they would not make it back to England. Fortunately the newly built emergency landing strip at Azeville was nearby and West, using all his skill, brought 'By Golly' in for a belly landing on the steel mat. As she ground to a halt the crew piled out and in a few seconds the plane was a mass of flames. The crew escaped without a scratch and were full of praise for their pilot. The landing strip was protected by an anti-aircraft unit and by one of those coincidences which even a writer of fiction would hesitate to use, one of the gunners was Capt. Milton Zola, brother of S/Sgt Harold Zola, the tail gunner of 'By Golly'. As they embraced, Harold greeted his brother with the words, 'Hi Milt – been waiting to drop in on you!'

Two days later they were back in England, showing their souvenirs – German helmets – to the other crews of the 598th BS. West and Zola were even interviewed for a radio programme, *The American Eagle in Britain*, which was scheduled to be broadcast on 5 August, by which time West would be dead and Zola would be in a POW camp.

Flying an unnamed brand-new B-26, with another nine-man crew, they took off on 1 August for another bridge over the Loire. This one was at Les Ponts de Ce, near Angers, and once again they were lead aircraft in the second box. Unfortunately the Luftwaffe fighters put in one of their increasingly rare appearances as the group approached the target and West, in the lead, was singled out for special attention. Cannon shells from Bf109s and FW190s exploded in the cockpit and a fire developed on the flight deck. As they began to lose altitude and drop out of the formation, Capt. West gave the order to abandon aircraft while he fought to hold the plane steady. The navigator, Lt Cramer, and Capt. West lost their lives in the crash, but the remainder of the crew escaped and spent the rest of the war in POW camps.

In a remarkable show of affection for their dead comrade, 598th Squadron personnel undertook the task of getting Capt. West's English spaniel Jiggs back home at the conclusion of the European war. Despite official setbacks and many difficulties, they eventually succeeded with their mission and Jiggs was safely delivered to Capt. West's two-year-old son Johnny in Sardis, Tennessee. The dog lived to the ripe old age of twenty, but tragically Johnny died in a flying accident in 1970, only a couple of years older than his hero father.

While assisting the author in research at Maxwell AFB Jack Stovall discovered that his cousin, Capt. West, had been awarded the DFC for his experience in crash-landing 'By Golly' without injury to the crew. West had died without knowing this and because of some oversight none of his family nor his widow were aware of the award. Jack Stovall wrote to the authorities and in 1978 the medal was finally handed over in the presence of West's family and US Army officials.

Poor visibility or heavy clouds in the target area were the cause of ten Rivenhall missions being 'scrubbed'. One of these was mission 63 on 7 July which was recalled because of bad weather but it did not prevent the group from receiving a severe mauling in the Laval area. Major Bronson, flying the lead position in the second box, was the only loss. He was using Lt-Col. Allen's aircraft 'Seawolf II' and Allen was not amused when he heard the news. Twenty-four aircraft were damaged by flak.

The 397th also served in a close support capacity for the Allied armies breaking out of the invasion beachhead and they were frequently in action against enemy troop concentrations, fuel and ammunition dumps. Mission 71 on 18 July saw them assisting the British Second Army by taking part in a massive bombing concentration around the historic town of Caen in a bid to break the deadlock between the opposing forces. The little village of Demouville was their target, some four miles East of Caen. Visibility was poor, probably not helped by the saturation bombing, and half the group were unable to bomb the primary target. Capt. Thompson's luck let him down once again and they were hit by flak, wounding Lt Stephenson, the bombardier. Despite their difficulties, 'Tommy' got them back again, as he did on so many occasions. One week later they performed a similar task for the US First Army when they bombed the area around Montreuil in support of Operation COBRA and the Allied sweep across France began.

Close formation flying was always an unnerving experience, particularly when there was cloud about. Two days before the group moved to Hurn, a tragic accident occurred when the aircraft piloted by Lt Gross hit the rudder of the lead aircraft in the second box, flown by Lt Garretson. Gross managed to keep control and eventually returned to base but the other aircraft went into a violent spin which prevented any of the luckless crew from escaping. Only six days later luck finally deserted Lt Gross when his plane, hit by flak near Dreux, rolled out of formation and disappeared into the cloud which covered the area.

Appropriately, the final mission from Rivenhall for the 397th was bridge at Epernon. Thirty-six aircraft in two boxes dropped a total of 200 1,000-lb bombs. They were led by Lt-Col. Dempster, who had the distinction of flying both the first and last missions from Rivenhall in 'Collect on Delivery', his personal plane. Two aircraft from the 598th BS, flown by Capt. Hawthorne and Lt Ryherd were lost on this mission.

During their stay the group flew 86 missions in 111 days and unloaded a total of over 4,500 tons of bombs on a variety of enemy targets. In the process they lost some sixteen aircraft – how many of the damaged ones made it back to be eventually repaired is not known.

The group continued to add to their reputation after they left Rivenhall. An attack in the morning of 23 December 1944 against a railway bridge at Eller in Germany earned them a Distinguished Unit Citation for outstanding performance. Despite the lack of fighter cover, the group successfully hit the bridge, but were immediately set upon by a large force of Bf109 fighters. During the subsequent fight, seven Marauders were lost, and of the planes which returned to base only five were undamaged.

By the time the war in Europe ended, the group had flown 239 missions, the final one taking place on 20 April 1945, exactly one year after the first Rivenhall mission. Coincidentally, it was also the birthday of the man who was indirectly responsible for the group's formation – Adolf Hitler.

With the departure of the American squadrons, Rivenhall settled down to a period of comparative calm, which was to last throughout the months of August and September. Then, on 1 October the peace and quiet of the surrounding countryside was once more shattered by the arrival of the final players in the drama being enacted on the East Anglian scene. Following the tenancy by the fighters and medium bombers of the US Army Air Force, the airfield now welcomed the Royal Air Force.

The Boys in Blue

Many of the airfields in the district were now lying idle and offered an ideal strategic position for attacking the enemy strongholds across the North Sea, and accordingly it was decided to transfer the RAF tugs and gliders of No. 38 Group from their bases in the Oxfordshire countryside. Nos 295 and 570 Squadrons' Stirlings towed their Horsa gliders from Harwell, arriving at Rivenhall on 1 October 1944 with the main party following at midday on 7 October.

Each squadron was composed of two flights: 295 were coded 8Z for 'A' Flight with propeller bosses painted red, and 8E for 'B' Flight, with yellow bosses. 570 Squadron codes were E7 and V8. The station strength was 251 officers, 603 NCOs and 1,838 other ranks. These totals included some 300 female personnel, housed in the 'WAAF Site' on the north-eastern perimeter of the airfield, in Storey's Wood. The R/T callsign for the control tower at Rivenhall was 'Snuggle' and for 295 aircraft it was 'Icepack'.

The Horsa gliders were only slightly smaller than the four-engined Stirlings, and with some sixty tugs and a glider apiece the airfield took on a distinctly crowded appearance. Many of the RAF personnel made the trip from Harwell to Rivenhall in the aircraft they serviced and, for most, it was their first experience of flying. L. A. C. Lowe was a flight mechanic (engine) in 6295 Servicing Echelon. His flight in a Horsa was an experience which he 'could not recommend for first time flyers. The large gap all round the door produced a noise like an express train going through a tunnel and the motion of the glider, bobbing up and down on the tow rope reminded me of a lift shooting up and down between floors, without stopping. The pilot did a tight turn over the runway and we landed in time for lunch – but none of our party felt like eating!' Jim Swale was an instrument technician with 6295 SE. He thought the whole lift was very precisely organised and well disciplined, much better than surface transport as the lads were able to take personal kit, toolboxes and bicycles with them in the gliders.

Originally formed in 1942 as a carrier squadron for airborne forces, 295 had the task of dropping pathfinder troops on three Dropping Zones (DZ) on the eastern flank of the invasion. Just before midnight on 5 June, six Albemarles took off from Harwell to set up radio beacons to be used as markers for the drop by the 6th Airborne Division, which occurred in the early hours of D-Day. 570 Squadron was formed in 1943 and many experienced crews from 295 were transferred to the new organisation. The exploits of these two squadrons during the Arnhem operation lies outside the scope of this chronicle, but again they contributed to the saga of that tragic but heroic operation.

Both the Stirlings and the Horsas were camouflaged in the standard RAF Night Bomber scheme, dark green and dark earth above and matt black undersides.

The Short Stirling Mark IV differed from the earlier marks in several external features. The nose machine-gun turret had been replaced with a clear perspex observation position and the mid-upper turret was removed and faired over. This reduced the defensive armament to the rear turret mounting four Browning .303 machine guns. The bomb bays were retained and each aircraft was equipped with a 'U' shaped tow line attachment under the tail turret. An exit hatch in the rear of the fuselage was used both by the paratroopers and also for dropping supplies, which were contained in a large wicker hamper, mainly used for packing the jerry cans filled with high octane fuel. The parachutes were fixed on a static line and two army personnel were responsible for the actual drop, on a given signal.

The requirements for supply dropping were sometimes altered from hour to hour, which necessitated changing the containers many times as the air force struggled to keep up with the needs of the ground troops. Occasionally, this involved a complete change to the payload of containers, lined up in the huge bomb bay, which was over 42 feet long.

The Stirling had seven fuel tanks in each wing, four main tanks and three others which were only used for the long trips to Norway. If all tanks were in use it seriously reduced the amount of payload which could be carried and if for any reason an aircraft had to return early, the excess fuel had to be jettisoned before landing, as the aircraft would be too heavy with all tanks full. The four Hercules XVI two-row radial engines had to have a daily inspection and the mechanics required a good head for heights as they were 15 feet off the ground. If the plugs had to be changed it meant a total of 112 plugs, two for each of the fourteen cylinders on each of the four engines, not the most pleasant of jobs on a cold day with a biting wind cutting across the airfield. In icy weather, the mechanics had quite a job to get up onto the wings, which had a pronounced slope. Safety mats were supposed to be used when working on the wing, but few were available at Rivenhall. After every trip each of the engines had to receive attention, topping up with oil to the maximum of 27 gallons and at the end of the day – or night – the covers had to be secured onto the engines ready for the next flight.

The Stirling suffered from its original 1936 specification, which had dictated a span of 99 feet to enable it to be housed in existing hangars and this seriously reduced its ceiling. Lancaster crews were secretly glad to have Stirlings along on the nightly raids, thinking that they would tend to receive the unwelcome attentions of the flak gunners and give the Lancasters, flying several thousand feet higher, an easy ride. The records tend to support this view. The Stirling crews seem to have had a soft spot for their aircraft despite its shortcomings. Loyal crews maintained that they were 'built like a railway carriage' and could absorb a terrific amount of damage – which they frequently had to do when flying in the low position with

Main Force bombers. Stirlings at 14,000 feet were rare according to official accounts.

The stalky undercarriage was a distinctive feature of the aircraft. In order to improve the Stirling's handling in take-off and landing, it had been decided to increase the ground angle of attack of the wing. As manufacture was too far advanced to change the angle of incidence of the wing relative to the fuselage, this could only be done by lengthening the undercarriage. The resulting structure had an unfortunate tendency to collapse with any sideways movement or when turning off the runway too fast. In addition, the tyres were prone to bursting after picking up flints or stones from the runways or perimeter track.

The Airspeed Horsa was perhaps the most successful glider used by any of the airborne forces in World War II. It was a high wing monoplane with a span of 88 feet (only 11 feet less than the Stirling) and extremely large split flaps, which have been described as 'being as big as barn doors'. These flaps were the secret of its very steep descent and short landing run. A standard Horsa comprised thirty production units, mainly produced in furniture factories. Hundreds were turned out by Harris Lebus and a few by the Austin Motor Company and Airspeed at Christchurch. It weighed a total of 7,000 lbs and was capable of carrying twenty-nine fully equipped troops or a 75-mm pack howitzer. From the outside the Horsa looked like any other large aeroplane, but without engines. As soon as one stepped inside the difference was evident – it had the appearance of a big model, everything in the structure being of wood. The sound of footsteps and the echoes in the fuselage gave the impression of being in a wooden hut, and seated at the dual controls it came as something of a surprise to find that they were made from laminations of multiply. In flight the controls were heavy, as one would expect from such a large machine and the ailerons had a 'delayed action' effect. After applying aileron to bank the aircraft in a turn or to pick up a dropped wing, there was no immediate reaction, which invited the application of more ailerons. When it took effect and the ailerons were centralised, the wing kept on lifting, requiring opposite aileron, and so on. The result was over-controlling until the pilot got the feel of it.

Contrary to popular belief, glider pilots were not always volunteers. Such volunteers as there were fell far short of the requirements and in order to offset the heavy losses incurred during the Arnhem operation, many RAF pilots were seconded to the Glider Regiment. This enforced transfer was the reason for a certain amount of bitterness in some glider pilots, who felt they had been trained for bigger and better things than gliders.

A training programme was started to give the tug and glider crews further experience of this most difficult of air manoeuvres but the weather prevented much being accomplished. During the month of October operations were flown both by day and (more often) at night. This was to be the pattern for 295 and 570 Sqdns for the coming months, with glider-towing exercises taking part during the day, involving a boring three-hour glider tow and a further one-hour cross country after the glider had cast off. An indication of the prevailing weather can be learned from the fact that one exercise (ESSEX) was postponed for four consecutive days; on 29 October it took place in low cloud and rain.

A tactical landing was a heart-stopping sight. The Horsa, travelling at 100 mph, would cast off from its tow, lower its huge flaps and point its nose downwards into the wind. This gave a descent slope of 1 in 1½ and to the onlooker seemed certain to end in disaster! At the last moment the glider would round out, landing on its main wheels, rock onto its nose wheel and apply brakes. This manoeuvre, if correctly executed, could enable a skilful pilot to land the Horsa in only a few lengths. The technique for mass landings was rather different; much longer landing runs were used to enable gliders to reach their appointed places and leave the area clear for others.

As might be expected, in the early stages of training many gliders landed outside the airfield perimeter, often in a badly damaged condition. The news of a crashed glider would flash around the district, and unless a guard was placed on the wreck, many parts of it would swiftly disappear. Plywood was a rare commodity during the war years and a Horsa was built of very little else. Bungee rubber was also used in its construction and this also was quickly spirited away. The author well remembers acquiring $^1/_{16}$-inch thick plywood, useful for making model aeroplanes, which was removed from a Horsa, stuck firmly in a hedge. Even today parts of fuselages may still be seen, used as sheds in gardens or allotments.

Rivenhall airfield was occasionally a welcome sight for aircraft other than an its own squadrons, particularly those in distress. On 30 October 1944, two pilots chose the Rivenhall runways to put down their damaged planes, the first a B-17 Fortress of the US 8th Air Force, based at Podington, which had been hit by flak and landed at 15.10 hours. The damage to the hydraulic system was repaired and the aircraft took off next morning, only to crash into the river at Bradwell, from unknown causes. The second incident was at 23.45 when a Halifax from 427 Squadron (RCAF) based at Leeming returned from a night raid on Cologne, also suffering from flak or night-fighter damage. In a previous incident a Main Force Lancaster flown by F/Lt F. G. Williams was hit by flak on the night of 12 July when taking part in an attack against the marshalling yards at Vaire. Both starboard engines were out of action, the undercarriage was down, the instrument panel shattered and one of the 500-lb bombs refused to drop. Despite all these setbacks F/Lt Williams nursed his shattered aircraft back to England and got down safely on the flare-lit runway at Rivenhall.

Five aircraft were detailed for a tactical bombing exercise on 4 February 1945, which did not get off to a very auspicious beginning. Waiting on the end of the runway was a private car being driven by George Brown of Braintree. The night was dark and George, thinking he had got a 'green' from the control van, proceeded on his way, only to collide with the tail of a Stirling which was taxiing into position – he had mistaken the starboard navigation light for the OK from control. The remaining Stirlings took off but one – LJ995 'The Bushwacker' piloted by F/Sgt Halford – crashed soon after take-off at Lanham Green with its load of twenty-four 500-lb bombs which produced a noisy pyrotechnical display as they exploded. The crew scrambled clear and received only minor injuries. The crash of 'The Bushwacker' soon brought the local police force to the scene, one of whom was PC Chalkley, the Silver End 'bobby'. He was the source of terror to small boys in the neighbourhood, who had learned to fear his unexpected arrival on his upright police bike whenever they were engaged in illegal activities such as 'scrumping' apples or finding their way onto the airfield through gaps in the perimeter hedge. When he arrived at Lanham Green to assist with the evacuation of the inhabitants from the nearby cottages, he was asked to return by one old lady who had forgotten her false teeth, and he did!

Many local residents near the airfield still recall the near disasters and accidents which they witnessed during the war years. In the majority of cases, the crashes, although frightening to aircrew and onlookers, were rarely fatal. On the night of 3 December 1944, F/Sgt Drife, flying Stirling LK273 E7-X, overshot the runway and tore off a wing with the result that the aircraft burst into flames. The crew scrambled clear and there were no injuries.

Although many preliminary plans were made, glider operations tended to be few and far between and the Stirlings were used extensively for several other duties besides that of towing Horsas.

The full story of the SOE (Special Operations Executive) may never be known. Its job was to organise sabotage and guerrilla warfare in the occupied countries from its headquarters in Baker Street, London. Hardly anything is known of the organisation and many of its records were burned at the end of the war, reputedly on the orders of high authority. The part played in these secret operations by 295 and 570 Squadrons was the supply of arms and equipment to the many resistance groups in occupied Europe and sometimes the transportation of agents. The navigational efficiency required, in order to locate a pinpoint in the heart of enemy-occupied territory, had to be of the highest order. In addition, each aircraft, operating alone, had to ensure that the drops were made from a precise height with the greatest accuracy. The correct signal had also to be received from the reception committee, whose members often waited for the supplies at the risk of their lives.

The Norwegians were reported to be among the best at marking the dropping zones and in giving the correct signal from the ground.

Of necessity, aircraft flying to Norway had only approximately half an hour in which to locate the zone, in a land of mountains, not noted for good weather, before a shortage of fuel made their return essential. The total flying time was approximately nine hours.

On Thursday 2 November 1944, the first attempt by 295 Squadron to supply the Norwegian resistance movement at night was made. Eight aircraft took off in good weather but met 10/10 cloud cover over Norway which forced them up to 9,000 feet. G/Capt. W. E. Surplice, DFC, DSO, the Station Commander, was piloting the 8th aircraft LK171 in the role of observer. The operation was not successful, only one of the Stirlings dropped supplies and that on an alternative DZ. G/Capt. Surplice's aircraft (coded WES) was the only one that failed to return, but faint signals from a dinghy were thought to hold some chance of rescue. On the next day, six aircraft took part in a search of that area of the North Sea where the signals were picked up and although they flew to within 30 miles of the Norwegian coast no dinghy or wreckage was sighted. Nevertheless, a further search was made on the following day but the results were the same. The loss of this popular officer was felt very keenly by the squadron and station personnel. G/Capt. Pope, DSO, took over command on 11 November 1944.

Jim Swale of 295 Sqdn 'B' Flight recalled that he sometimes met the agents as they prepared to board the Stirling. One attractive girl shared a small bottle of brandy with the ground crew on a cold night, while waiting for take-off. They often waited about on these missions until the aircraft returned, to find out if the agents had been able to drop on the DZ.

Supply dropping at night was nearly always a chancy affair with only a limited success rate. A number of agents who should have parachuted into occupied Europe were brought back to Rivenhall because of uncertainty over the DZ. Sometimes they were unable to jump because of a faulty parachute harness. On 18 November 1944, nine aircraft attempted to drop supplies to the Dutch resistance. The weather was bad, visibility low and the night was dark. No lights were visible on the DZ and the pilots had no alternative but to return with their loads, a familiar pattern for many SOE missions. On this occasion it was a blessing in disguise. We now know that the German Military Intelligence (the Abwehr) had thoroughly penetrated the resistance groups in Holland and much of the arms and equipment were being delivered straight into German hands.

The month of January 1945 saw all the vagaries of an English winter; snow, frost and fog all combined to severely curtail all flying activity. Instead, aircrew listened to lectures on such entertaining topics as 'Estimates of Enemy Ground Strength and Disposition in the Western Battle Area' or watched films such as 'The Battle of Britain' – not the 1970 Cinemascope version! In desperation an impromptu cross-country race was organised to keep the lads busy and even chopping wood became a diversion. When the weather allowed, training continued; container dropping was practised with the aid of radar, for which the station at Oxford seems to have been used and Rebecca bombing on the range at Wainfleet was also carried out. No operations were flown by 570 Squadron either by day or night during January but on 1 February five aircraft bombed a target in Germany and one SOE mission to Holland was flown. This pattern continued for several nights in an endeavour to make up the lost time. The operations flown during this period went under the code names 'Rummy', 'Draughts', 'Bluebottle', 'Black Widow' and 'Dudley'.

A note of satisfaction is revealed in the squadron record book concerning a night operation on 24 February when LK144, a Stirling of 295 Squadron piloted by S/Ldr Stewart, dropped eleven containers on a DZ inside Germany, the first time SOE drops had been carried out inside the Third Reich. The drop was made in good visibility, assisted by a diversionary bombing attack on the railway at Herborn from 5,000 feet.

Warrant Officer Peter Baldock, of Braintree, flew with 295 Squadron as navigator, most frequently with B-Beer LK129. He recalls that SOE drops were made whenever the weather conditions were favourable. During the month of April 1945 the aircraft of 295 and 570 Squadrons flew SOE missions to aid the resistance fighters near Copenhagen and other places in Denmark and Holland. Wherever possible these operations were carried out in at least half- moon conditions, as the final part of the navigation had to be visual. Returning from a successful drop to the Freedom Fighters in Holland during this period, W/O Baldock recalled they were coned by the searchlight defences. The pilot pushed the control column forward to build up speed and began to take evasive action from the anti-aircraft fire which was being hurled at them. As they levelled out from the high speed dive, in which the aircraft touched over 300 mph, the rear gunner got in a good burst and knocked out one of the searchlights.

On the night of 26 April seven aircraft took part in an SOE operation code-named 'Table Jam' and were assigned to different targets in Denmark. The defences were reputed to be severe and the instructions were to 'go in low'. B-Beer returned safely, after disposing of its supply load of twenty-four containers and two packages, but the crew reported seeing one of the Stirlings, which was flying extremely low, belly-in on the shores of an inland lake. Two aircraft were lost from 295 Squadron: LK567 piloted by F/O Griffiths and LJ950 piloted by W/O Dax, whose luck finally deserted him. He piloted LJ995, the Stirling which had crashed on 4 February as related earlier. One of the hazards of low flying was small arms fire; the wireless operator of a 570 Squadron Stirling on the night of 23 April was hit in the shoulder by a spent bullet, which was assumed to have been from small-arms fire.

The Short Stirling was the first four-engined bomber of the war to be used by the RAF, although its limited ceiling prevented it from ever attaining the fame of its later compatriots, the Halifax and the Lancaster. Nevertheless, it did continue to serve in its original capacity with 295 and 570 Squadrons, both of which underwent bombing training in addition to their other varied duties. In February 1945 the crews put their training to good use when they were called on to perform tactical bombing in support of the 21st Army Group. These operations called for close support bombing, at night, just behind the front line. The aircraft successfully delivered their loads of twenty-four 500-lb bombs, bombing being accomplished using the 'Gee' radar homing system. Under optimal conditions using this system it was possible to bomb within an accuracy of 200 yards. 'Gee' had been used to lead the largest attack so far mounted by the RAF when, on 30 May 1942, a force of over 1,000 aircraft bombed Cologne. It was arguably the most successful raid of the war up to that time, largely thanks to 'Gee', which also proved useful in guiding homecoming bombers back to base, particularly in bad weather.

The Rhine Crossing

Both of the Rivenhall squadrons, together with others of Nos 38 and 46 Groups, took part in Operation VARSITY, the final large-scale airborne operation of the war. On Saturday 24 March 1945, in conjunction with Allied troops on the ground, paratroops and gliders of the American 17th and the British 6th Airborne Divisions were landed on the east bank of the Rhine near the town of Wesel, which lay just over the Dutch/German border. Interestingly, Wesel had figured in airborne planning possibilities as early as August 1944. Two days beforehand the airfield was 'sealed off' and no one was allowed to leave, post letters or use the telephone.

Mr F. V. Foreman, who was at Rivenhall when it was under construction, was now serving king and country by unloading bombs from the railway sidings at nearby Kelvedon, on to the waiting Rivenhall lorries.

My home in those days was at Buckhurst Hill in Essex and, with the help of my former Site Agent who was still engaged on maintenance at Rivenhall, I used to slip home in his car – using the contractors entrance. I almost 'had my chips' one Sunday afternoon when I returned from such a trip as I found the camp was sealed off and barbed wire was everywhere. Fortunately I knew of a convenient hole in the hedge so I soon got back on the inside. This was the eve of Operation Varsity

– one of the worst kept secrets of the war. It was certainly common knowledge in our local hostelry, The White Hart in Kelvedon.

On 23 March the glider pilots, aircrews and airborne troops were carefully briefed. They were told that the success of their mission would considerably hasten final victory. At 2.00 a.m. on the morning of the 24th the crews were called, given breakfast of bacon and eggs, and received a last briefing which included final weather reports. The Stirlings were marshalled along the perimeter track on either side of the main runway and the Horsas loaded with troops and equipment were parked in a tight group at the end of the runway. It must have been an inspiring sight as 121 aircraft (not counting the spare tugs) prepared for take-off. What a pity no one had the foresight (or time!) to take a few photographs of this unique occasion. Luckily someone did on the next operation (DOOMSDAY) and by reference to that photograph we can visualise the scene in the early light as the mass of Stirlings, moving nose to tail, slowly taxied into position. The noise from the 244 Hercules engines must have been deafening.

Gliders were lined up, roped to the tugs and parked awaiting take-off. The Control Officer signalled the first aircraft away at 07.00 hours. As one Stirling cleared the end of the runway another was given the signal to go. The take-off, with a fully loaded glider, was sluggish. The Horsa was first off the ground, after a run of about thirty seconds, and three seconds later the Stirling became airborne.

A total of sixty-one Stirlings towing sixty Horsas took off, one Stirling carrying press observers. Three tow ropes broke, which necessitated the spare tugs being called up and resulted in late take-offs for these serials. In the meantime the Rivenhall squadrons flew on to rendezvous with the others of 38 and 46 Groups over Hawkinge in Kent. The four-hour flight, in perfect spring sunshine, was uneventful except for the airsickness experienced by the troops and passengers in the gliders. This was more severe than usual due to the turbulence caused by the 'prop wash' from the hundreds of aircraft in the column. Many glider pilots and troops were even looking forward to the landings as a relief from the airsickness! (A note in the station record book provides an interesting comment, 'Air sickness is a rare occurrence in a Stirling.' It would be nice to know whether this had any basis in fact.)

The whole convoy consisted of 1,696 transport planes and 1,348 gliders (some in double tow) carrying 21,680 troops, with 889 fighters acting as escort. In addition, a further 2,153 aircraft were engaged in silencing the defences in the target area or ranged far into Germany in order to prevent the Luftwaffe from interfering with the Allied landings. As if this display of Allied air superiority was insufficient, 2,596 heavy bombers and 821 medium bombers attacked airfields, bridges, marshalling yards and any other suitable targets. The sky train took two and a half hours to pass a given point and the thunder of their engines coming out of the west was a welcome sound for the Allied ground troops on the banks of the Rhine.

The Rivenhall squadrons had the distinction of leading the airborne fleet over the Rhine at Wesel, thus living up to the motto of 570 Squadron, 'We Launch the Spearhead', although 295 Squadron's 'Help from the Skies' could not be considered inappropriate.

The drop was not without incident. Fighter Bombers of the 2nd Tactical Air Force, whose task it was to nullify the defences, had been partially prevented from doing so by the early arrival, by seven minutes, of the tugs and gliders. The paratroopers and gliders had to descend through a cloud of dust and smoke blown by a westerly wind across the landing zones and in the face of anti-aircraft fire and small-arms fire from the ground defences. The British glider landings were remarkably accurate, many touching down within 20–30 yards of their objective. Fortunately, the casualties were light despite the fact that less than a quarter of the gliders were undamaged. The Glider Pilot Regiment casualties were 38 dead, 37 wounded and 135 missing, the majority of these being eventually accounted for.

Sgt Tony Wadley was one of the ex-RAF pilots seconded to the Glider Pilot Regiment in the autumn of 1944 who flew on Operation VARSITY as second pilot to Sgt 'Tug' Wilson from Doncaster. His recollections are probably typical for many who flew this last great airborne operation of the war:

Each glider crew received a low-level photograph of their landing zone at the briefing, close to a farm house near the German town of Hamminkeln, which was to be used as Divisional HQ. We were woken very early on Saturday 24 March with tea brought to our hut by WAAF cookhouse staff. Our Horsa was loaded with a Jeep and trailer carrying radio equipment together with five troopers from the Devon regiment. Take-off was soon after seven o'clock, but as we gathered speed down the runway our tow rope broke before we were airborne. A waiting tractor towed us clear of the runway and we were eventually connected to LK351 E7-N, a 'spare' Stirling of 570 Sqdn piloted by F/Sgt Sabourin. We finally took off at 8.55 a.m. long after the remainder of the Rivenhall serials had departed.

The journey took three and a half hours and the only incident I recall was the sight of a Horsa a little to our right breaking its tow over the North Sea; the Air Sea Rescue boys no doubt speedily picked them up. When we reached the Rhine we could see smoke rising from the ruins of Wesel and as we passed over the river a bit of flak started to come up. The release point came and we cast off and headed towards the field. Our touchdown was smooth enough but slightly too far into the LZ as we realised when we rapidly approached the end of the field and a low embankment. It was obvious that even with the 'barn door' flaps down and brakes hard on we had no hope of stopping in time. With an almighty crash the undercarriage was wiped off by the embankment and the nose ended up in the ditch on the other side with the rest of the aircraft pointing towards the sky behind us. Everyone climbed out somewhat dazed but unhurt and we took stock of the situation. Normally we would have swung the nose section open, laid down our two portable ramps and driven the load out of the glider – that is how it had been done on exercise! 'Tug' Wilson offered to place some plastic explosive round the hinges of the nose section and we all retired behind the embankment to await results. Nothing happened so 'Tug', with a show of commendable courage went back and fitted some new detonators and this did the trick. After a good deal of effort on all our parts we got the jeep and trailer out, the troops on board and off they went down the road. Our Horsa, or what was left of it, was forming an effective road block which required the assistance of a passing Bren carrier before the road was made passable. In the process large pieces – the whole tail unit for example – came adrift and anyone passing later must have thought we had made a singularly bad landing.

Our group of glider pilots were allotted the task of providing defence for Div. HQ and each crew dug themselves into foxholes in the small orchard by the farmhouse and awaited developments. As it turned out, we didn't need to do any 'defending' as it was fairly peaceful in our area. The weather was fine and sunny and we spent the time watching the 2nd TAF Typhoons peeling off above us and diving down to fire their rockets at an unseen enemy to the east of us. We spent three nights by the farmhouse before being ordered to return. On 29 March we were flown back to England and landed at Down Ampney in Gloucestershire.

Wadley was able to return to powered flying at the end of 1945 and finally left the RAF in 1950.

Several aircraft were hit by light flak, among them a 295 Sqdn Stirling piloted by W/O Symons, 8E-J, whose port inner was hit and set on fire. W/O Symons gallantly held the Stirling steady while his crew bailed out and left it too late to make a successful escape himself, being killed by striking the tow hook. The aircraft was seen to crash in flames. Two of his crew reported back to Rivenhall a few days later. About 300 gliders were damaged, many of them severely, and ten were shot down. Among them was the Horsa towed by B-Beer, one of the first to cross the Rhine.

The Stirlings returned from the Rhine dropping zones in time for lunch after a five-hour flight. A common feature reported by many crews at the debriefing (or interrogation as it was previously called) was the sighting of a V2 rocket climbing from its launch pad in Germany, leaving behind its characteristic trail of white vapour. The following day all crews were briefed for possible re-supply dropping to the Allied troops on the Rhine, but in the event no sorties were flown and the crews were eventually stood down.

On 7 April a large-scale operation, coded 'Amhurst', took place when eight aircraft from 295 Sqdn each dropped fifteen French troops of the 2nd SAS plus four containers on to a DZ in Holland to the east of the Zuyder Zee. On this occasion, as on many others, navigation was by Gee. With the Allied armies racing across Europe and the main Channel ports still in enemy hands, the question of supplying essential materials to the forward areas became of paramount importance. During the month of April, Stirlings from 295 and 570 squadrons were called upon to ferry high-octane fuel, in 125 five-gallon jerry cans, to the Allied fighter airfields. It says something for the logistics of the war at that period, when it is realised that in order to carry a payload of 625 gallons to the waiting fighters, each Stirling used between 1,000 and 1,200 gallons of fuel on the round trip.

As the war in Europe drew to its close, the Rivenhall Stirlings were used in yet another capacity, as personnel carrier aircraft for Operation USEFUL. The advancing Allied armies liberated many POW camps and the freed prisoners were transported by road to Brussels. There, the Stirlings picked them up and flew them back to England, each aircraft carrying twenty-six released POWs which offset the poor economics of the fuel-delivery flights referred to earlier. In the months that followed, Rivenhall became one of the busiest airfields in East Anglia, so much so that eventually customs facilities were installed. The Stirlings continued to ferry the returning ex-POWs and servicemen, on leave for demobilisation, during the weeks following the end of the war in Europe.

8 May 1945 was VE (Victory in Europe) Day and all station personnel were paraded in No. 1 Hangar and addressed by Group Captain Pope, DSO, the Commanding Officer, who read first a roll call of casualties and then the statement by the Prime Minister announcing the cessation of hostilities in Europe. The station was then placed on 'stand down' until 14.00 hours on the 9th. On this day all crews were briefed for Operation DOOMSDAY, a forbidding code name in a war which had seemed to consist of endless code names. This one involved the squadrons of No. 38 Group airlifting troops for occupation duties in Norway. Following a false start on 10 May when the take-off was suddenly cancelled due to an active front over the North Sea, twenty-three aircraft from 295 and

twenty-four from 570 took off on the 11th after a further four-hour postponement.

Tragically, three aircraft from No. 38 Group crashed due to the bad weather. Among those lost was the AOC of No. 38 Group, Air Vice Marshal Scarlett-Streatfield, together with many soldiers of the 1st Airborne Division.

The manifest for W/O Baldock's aircraft LJ573 8Z-Q included sixteen RAMC personnel, one motorcycle and one dog. An ex-Luftwaffe airfield, Gardermoen, 40 miles north of Oslo, was the destination, where the German commander surrendered to the senior British officer. The German personnel continued to man the airfield services and to assist the turnaround of aircraft.

The Stirling was not the best shelter when it rained. The author recalls being caught in a severe downpour and thankfully scrambling into a handy Stirling parked near the edge of the airfield, only to find there was quite a bit of rain seeping in. On a visit to Gardermoen in May 1945, Peter Baldock and the rest of the crew spent an uncomfortable night in their aircraft, using parachutes for bedding. It rained heavily and they spent much of the night trying unsuccessfully to find a dry place to sleep.

The station records for the summer of 1945 reveal an air of light-heartedness in sharp contrast to the seriousness of six years of war, with sports practices culminating in a station sports day at Witham on the 16 June (295 Squadron won handsomely) and about this time the first references appear of airmen being sent to the demobilisation centres, en route for 'civvy street'. The famous novelist Dorothy L. Sayers, who lived in Witham, gave a talk in the station cinema entitled 'The Playwright and the Theatre' on 27 June and on 3 July polling booths were erected in all parts of the camp. The 1945 general election, which gave a landslide victory to the Labour Party, was perhaps not much influenced by the 530 votes placed in the boxes at RAF Rivenhall, and a note of disgruntlement is revealed in the fact that 20 per cent of voters did not receive their ballot papers. Over 1,701,000 service men and women voted by post or by proxy to give Labour a total of 393 seats in the House of Commons.

At the end of the war many of the ground crew were treated to a flight over the shattered cities of Germany. Jim Swale flew in 'his' aircraft 8E-Q (seated on a wheel cover – no seats were provided) and was amazed at the destruction he saw in Essen, Dusseldorf and Cologne.

Despite the frivolities of post-war Britain, the squadrons maintained and attempted to improve their operational standards and mass glider lift practices continued. With what degree of determination is not known; perhaps they preferred the trips such as the one which involved transporting ENSA party concerts to the Army of Occupation in Germany and returning with the Liverpool football team, as happened to W/Cdr Angell, DFC, the OC 295 Squadron on 31 July. W/O Baldock is quite emphatic: 'Those were the best days' and it is easy to see why.

The war in the Far East ended with the Japanese surrender and VJ Day was announced for 14 August 1945. Now at last it was all over and the celebrations could commence. The service men and women, station staff, ground and aircrews attended the many victory celebrations held both on the airfield and in the surrounding towns and villages. A dance in the NAAFI went on until 03.30 hours; a stern note is sounded in the station operations. book that 'everything was kept well under control and the Fire Piquet patrolled until 04.00 hours'. In the neighbouring village hall at Silver End the local RAFA sponsored a Grand Victory Dance and Cabaret on the following night.

Later in the month preparations began for some long-distance trips to India and the Middle East, assisting Transport Command in ferrying spares and machine parts to Karachi and Cairo via Tripoli. For a few days there were no flying activities at Rivenhall; instead, all crews were nursing stiff arms and pained expressions following the series of inoculations.

The flights which began in September were made using the 'slip crew' system, whereby a crew took an aircraft partway along the route, handing over to another crew who would complete the mission. In this way the whole operation was speeded up and a quick turnaround accomplished. The loads were of a varied nature. W/O Baldock was unfortunate enough to be part of the crew when the aircraft was carrying a load of carrier pigeons back from Tripoli. He still recalls the smell!

During the month of September the squadrons provided aircraft for display at the Battle of Britain celebrations held at several RAF stations. Stirlings and Horsas flew to North Luffenham, Odiham, Tilstock and Finningley on the 15th, staying overnight before returning.

Both 295 and 570 Squadrons continued to carry out daily trips to airfields all over Europe including Brussels, Munster and Schleswig, taking mail and newspapers to the Army of Occupation. Occasional trips were also made to Copenhagen, Prague and Vienna but in addition to the forces mail and newspapers the Stirlings were often carrying clothing to the people of the occupied countries who had only recently been freed from Nazi bondage.

On 23 September a memorial flight was made to Arnhem, the scene of one of the bitterest actions of the war in 1944. It provided sombre memories for the crews taking part. Both squadrons had been involved in Operation MARKET GARDEN, a daring scheme to land paratroopers and gliders to form a corridor through enemy-held territory along which the army could smash their way into Germany, which took place only a few days before the move to Rivenhall. The key to the whole operation lay in securing and holding the bridge across the Rhine at Arnhem and as history now shows, despite the gallant sacrifices by the 1st Airborne Division, the bridge proved impossible to hold with the limited troops and supplies available.

Almost the last entries in the station operations book record the success of the RAF Educational and Vocational Training Scheme (EVT) in helping the men and women in their transition from service to civilian life. Classes in various activities were under way and a woodwork room was to be ready in January. It was destined never to be used.

Rivenhall continued to be one of the busiest airfields in the UK until January 1946 when Nos 295 and 570 Squadrons moved to Shepherds Grove airfield, situated 12 miles north-east of Bury St Edmunds. On 8 January 1946, 570 Squadron was disbanded, followed by 295 Squadron on 14 January. With their departure, Rivenhall ceased to be an operational airfield.

Post-War Development

With the end of hostilities, the governments of the Western Powers were faced with the problem of what to do with the millions of displaced persons from all over Europe. The upheaval of vast numbers of Poles, Czechs and Slavs escaping from the fury of battle or enslavement by the Nazis had resulted in an enormous post-war problem. A decision was taken to group the various nationalities together in camps and in the local area this resulted in the disused buildings of Rivenhall being taken over and turned into 'The Polish Camp'. Polish Army personnel, released from prison camps on the Continent, arranged themselves into a strong community at Rivenhall ,where they were joined by their families and other Polish servicemen who were being demobilised from the armed forces.

Over the years the Poles gradually became integrated into the East Anglian scene, marrying local girls and working alongside Essex workers in fields and factories. Some families emigrated to the colonies with government assistance and a number returned to Poland. By the mid-1950s the process of integration had been so successful that the need for the Polish Camp disappeared entirely and it was finally closed down.

A bold social experiment took place in the early 1950s when the Essex County Council set up the Wayfarers' Hostel in the old station headquarters buildings at Rivenhall. This venture was intended to provide a semi-permanent base for the itinerant travellers in the area and appears to have operated successfully for about five to six years. Not much is known or written about concerning these years, but no doubt a future chronicler would find it an interesting subject for further research.

During this time the giant electronics firm of Marconi Ltd had been looking around for a suitable site to expand their rapidly growing business in the field of communication systems. They first leased part of the airfield in 1956, gradually expanding over the years until by 1975 they had taken over all the old wartime buildings and the two hangars. The much vaunted 'bush telegraph' system, used by the gentlemen of the road, appears to have badly malfunctioned during this period, as many of them continued to turn up at the Wayfarers' Hostel long after it had been turned into a stores site by Marconi.

Rivenhall has occasionally been used since the Stirlings departed. Sometime in the 1960s a Super Sabre F100 made an emergency landing on the only existing runway; fortunately it was the main one and still in fairly good condition. The pilot took off successfully later in the day.

The airfield's two short runways and the majority of the perimeter track and hardstandings were broken up and used in the extensive road-building programme which took place in the 1960s, and the land returned to agricultural use.

The land which surrounded the airfield continued to be farmed by the various owners throughout the war with occasional help from the service personnel in their off-duty moments. Manley Nelson was an instrument repair mechanic with the 599 BS and was a frequent visitor to Woodhouse Farm, where his help was greatly appreciated by John Ambrose and his wife in the summer of 1944. The farm had been reduced from 270 to 50 acres by the building of the airfield. In 1980 when the author interviewed John Ambrose, he and his wife were just preparing to leave the farm which had been their home for fifty-five years. Still living with them was Florrie, who had come to them in 1941 as a member of the Women's Land Army at the age of twenty-two. The girls were another familiar sight in wartime Britain with their dark-green sweaters, brown corduroy trousers and wide-brimmed hats. Surely there cannot be many Land Army girls with a continuous service record of almost forty years! Manley Nelson now farms in Montana and, like farmers everywhere, complains at the price he gets for his wheat. He keeps in contact with friends in Kelvedon and still recalls the times he rode the tractor on Woodhouse Farm. Peter Baldock, a navigator with 295 Sqdn RAF, also helped at harvest times in 1945.

Scattered around the farm today are the rusting remains of wire mesh reinforcing matting, used originally for parking aircraft away from the concrete hardstanding, now serving as useful fencing. Metal stands for servicing the aircraft and much other paraphernalia whose original purposes are obscure, are also still in evidence. Lying on a pile of rubbish and half-covered with nettles, lies the metal post which once held aloft the windsock beside the control tower, long since demolished.

Anyone who had been stationed at Rivenhall would have great difficulty in locating the sleeping sites, mess halls or flight sheds, for in the majority of cases the smaller buildings have all been pulled down and in their place dense undergrowth grows, with an increasing rabbit population.

During the phenomenal summer of 1976 the ponds and ditches surrounding the airfield became completely dried out, revealing for the first time for thirty years many interesting and potentially dangerous items. A team of Royal Engineers was called in to dispose of some very corroded smoke bombs, discovered by the author and Ken Fisher. Less dangerous finds were parts of a Stirling undercarriage door and pieces of parachute dropping equipment.

Rivenhall is perhaps unique among East Anglian airfields in that many of the wartime buildings have been preserved for use by the GEC-Marconi Company Limited. How much longer it will serve to remind us of the sacrifices made by British and American aircrews who flew from its runways is a matter of conjecture. If ever the question of a third London airport is revived we may yet see Rivenhall reinstated on the list of possible sites, as it was at the time of the Maplin decision in the early 1970s.

In the meantime, there is an unusual memorial, only discovered in the last few years. A visitor from the Royal Horticultural Society establishment at Wisley in Surrey noticed some interesting plants near the hangar on the eastern side of the airfield. He identified several trees as a species known as Florida Oak and a large bush as a type of willow with straight stems used by the Navajo Indians to make arrow shafts. Neither the oak nor the willow are native to these shores and one can only speculate on their arrival in these foreign surroundings. In any event they provide us with a nice link with the wartime years when Rivenhall played its part in the great fight for freedom.

1. 'Maggi's Drawers', a natural-finish Mustang P-51B of the 380th FS. (Photo via K. Fisher)

2. 13 March 1944. Brigadier General 'Pete' Quesada, commander of the IX Fighter Command, pins the Air Medal on Colonel Ulricson. (Photo J. Ulricson)

3. Lt Ed Pawlak in the cockpit of his P-51B, 'My Pal Snookie'. (Photo E. Pawlak)

4. Nice shot of 'My Pal Snookie' sporting D-Day stripes. (Photo E. Pawlak)

5. Major Culberson, the CO of the 381st FS, in the cockpit of his camouflaged P-51B, shortly after his first victory on 3 March 1944. (Photo S. Blake)

6. Lt Charles Reddig shot down a Bf109 on the 8 April mission. His Mustang was named 'Limited Service'. (Photo S. Blake)

7. Aerial view of Rivenhall. A total of forty-seven Marauders are on their dispersal points and two B-24 Liberators are receiving attention. Note the markings of the old field boundaries. (Photo J. Snow)

8. A seven-ton high-speed tractor stands beside the control tower. On the right are the unfinished brick sheds which housed nitrogen and hydrogen bottles used for weather balloons. The tower was demolished in the 1960s. (Photo J. Snow)

9. American transport vehicles were a feature of the Second World War and this shot of the motor pool at Rivenhall contains a representative selection of many types. (Photo J. Snow)

10. The Stars and Stripes flies over the 397th HQ at Rivenhall. In the 1950s this became 'the Wayfarers' Hostel', a home for the travellers of the road. (Photo J. Snow)

11. Little external change is evident in this 1975 photograph. (Photo B. A. Stait)

Opposite above: **12.** Staff of the 397th Bomb Group at Hunter Field, Georgia, 11 February 1944. Back row from left: Capt. Joseph A. D'Andrea (Special Services), Capt. William Rafkind (Communications), Maj. Robert L. McCollum (Group Surgeon), Maj. Kenneth C. Dempster (Operations Officer), Lt-Col. Rollin M. Winningham (Deputy CO), Col. Richard T. Coiner, Jr. (Commander), Maj. Franklin E. Ebeling (Executive Officer), Maj. Kenneth R. Majors (Adjutant), Maj. Thomas E. McLeod (Intelligence Officer), and Lt James M. Snow (Photo Officer). Front row from left: Capt. George D. Hughes (Operations), Lt John F. Haupt, Jr. (Ordnance), Lt Charles H. Schultz (Weather), Lt Henry C. Beck, Jr. (Photo Interpreter), Capt. Earl W. Udick (Group Navigator), Capt. Elton G. Morrow (Maintenance Officer), Capt. Claude S. Funderburk (Supply Officer), Capt. Clarence R. Comfort, Jr. (Chaplain), Lt Robert J. Wood, Capt. Fred E. Seale, Jr. (Armament Officer), Capt. James M. Lynch, Jr. (Controller), and Lt Deane Weinberg, Jr. (Personnel Officer). (Photo J. Snow)

13. Marauders of the 596th and 597th BS flying over the Essex village of Coggeshall in May 1944. (Photo J. Snow)

14. Colonel Coiner accepts the hand of an RAF officer (thought to be Wing Commander Martin of No. 3 Group, RAF Marks Hall) on the occasion of the 'handing over ceremony'. In the background there appears to be a British Army band. (Photo via Mrs. I. Daughton)

Right: **15.** The aircraft of the 397th BG are allocated their aerial cameras, prior to the next mission. The group numbered sixty-one aircraft at this time. (Photo J. Snow)

Below: **16.** The Aero Club shortly after completion in 1944. These buildings formed the Polish Camp site after the war. (Photo via C. de Coverley)

Above: **17.** Trees and hedges show the passage of the years but the buildings remain much the same today. (Photo B. A. Stait)

Left: **18.** 598 Squadron occupy the low positions as the group set off for the marshalling yards at St Ghislain near Mons on 26 April 1944. (Photo J. Snow)

19. The Group Operations Room at Rivenhall, with Major Earl 'Bud' Udick, the Group Navigator, using the telephone. (Photo J. Snow)

20. Major K. C. Dempster (standing extreme left) and crew of 'Collect on Delivery'. This aircraft flew on the first and the last Marauder mission from Rivenhall. (Photo J. Snow)

21. Lt-Col. Berkenkamp, CO of the 599 BS (standing second from left), with his aircraft and crew. On his left is 'Big' Bill Bond who took over the position of Group Bombardier from Snyder. (Photo J. Snow)

22. Another shot of 'Draggin' Lady' with Capt. Monte Stephensen (standing second from left) with a nine-man lead crew just before take-off. (Photo J. Snow)

23. Captain Moses Gatewood (centre) poses with his crew wearing flak suits and special helmets which permitted the use of earphones. (Photo J. Snow)

24. 'Mama Liz' of the 597th BS was lost on the ill-fated mission of 24 June 1944 when four of the group were shot down by flak during an attack on a Paris bridge. (Photo J. Snow)

25. 18 July 1944 saw the demise of X2-V when the left main tyre blew on take-off during training and the landing gear collapsed. The crew of three escaped safely. (Photo J. Snow)

26. Photographed from the control tower, the wreckage of X2-V sends a plume of smoke skywards. The shooting butts are just visible in the background. Note the airfield identification letters RL and the lights for night landing. (Photo J. Snow)

27. Peyton Magruder, designer of the Marauder, with three squadron commanders at Rivenhall on 22 July 1944. Left to right: Lt-Col. McCleod (596 BS), Magruder, Lt-Col. Allen (598 BS – wearing six clusters to his Air Medal) and Lt-Col. Berkenkamp (599 BS). The aircraft is Allen's 'Seawolf III'. (Photo J. Snow)

28. Lt Hamer of the 597th BS in his bomb aimer's compartment. He had the distinction of being the bombardier with the best CE (Circular Error) in the 397th BG. (Photo J. Snow)

29. Captain W. Smith (left) and crew of 'Susan J'. (Photo J. Snow)

30. Captain Ed (Rich) Richardson and crew of 'Little Peedoff'. Richardson is on the left and next to him is the co-pilot, Richard Hughes of Memphis. The man on the right has the 598 BS emblem on his jacket. (Photo J. Snow)

31. 'Walking the props' to clear the engine cylinders of oil, Captain 'Tommy' Thompson (holding parachute) studies the maps before setting off in 'Lassie Come Home' on her 37thirty-seventh mission. (Photo J. Snow)

32. Members of the 597th BS make use of one of the many bombs stacked around the dispersal points to speed up loading. The aircraft in the background is 42-96125 (pilot: Lt Marcel Gleis) one of the last camouflaged B-26s off the production line. (Photo J. Snow)

33. 'Bar Fly' 42-96171 with Lt Senart and crew. Nevin Price is front row, centre. Note the bombs and tent in background. (Photo J. Snow)

34. Major K.C. Dempster (rear, second from left) with a nine-man lead crew. F/O Breen (front, extreme right) was injured in a crash at Chipping Ongar on 8 June 1944. (Photo J. Snow)

35. Marauders of the 597th BS setting off for another mission. In the background are the poplar trees which stand beside Woodhouse Farm. (Photo J. Snow)

36. Turning off the taxi track on to the eastern end of the main runway, the pilots wait for the green light from the mobile control wagon. (Photo J. Snow)

Left: **37.** 'By Golly' 42-96138 with twenty-three missions to her credit. Standing left to right: Lt Budge, co-pilot; Lt Cramer bombardier (killed 1 August 1944); Capt. West, pilot (killed 1 August 1944); Lt Daoust, navigator. In front: S/Sgt Zola, gunner; T/Sgt Natanek, radio operator; S/Sgt Picklesimer, gunner; T/Sgt Robinson, crew chief. The black and white spaniel is Jiggs, their mascot. (Photo J. Snow)

Opposite below: **38.** The end of 'By Golly'. The aircraft bellied in on the emergency landing strip A7 at Azeville, Normandy. Capt. West (third from left with hand on the propeller) was awarded the DFC for his action during this mission. (Photo via Roger Freeman)

39. & 40. Two photographs showing the bombing and subsequent results of the 397th attack on the bridge at Le Ponts de Ce on 1 August 1944. (Photo Henry Beck Jr.)

41. 'Dee Feater' of the 596th BS flown by Lt-Col. R. M. McCleod over the deserted airfield at Birch, between Kelvedon and Marks Tey. The road at the bottom of the picture is the A12. (Photo Charles E. Brown)

42. Marauders of the 596th BS at dispersal in July 1944. 42-96144 X2-C in the background was lost on the Eller Bridge mission on 23 December 1944. Photo probably taken on 8 August. (Photo Charles E. Brown)

43. Close formation flying by 598th BS on an early morning mission. From left to right: 42-96156 U2-G; 132 U2-B (Lucky Star); 139 U2-D (Susan J); 129 U2-A (Seawolf II). (Photo J. Snow)

44. A poor-quality snapshot of 42-96175 6B-Y (the serial number conflicts with 6B-V, which bellied in at Rivenhall prior to D-Day) taken at Grange Hill, Coggeshall on 17 June 1944. The crew and householders escaped injury. (Photo via J. Spurgeon)

45. A fine study of a B-26C Marauder of the 598th BS in natural finish. (Photo J. Snow)

46. Stirling IV and Horsas parked near the duck pond at Allshot Farm in 1945. (Photo via C. de Coverley)

47. One for the album! Ground crew in front of 'Goofy' 8E-O in the spring of 1945. In the original picture a total of twenty-five unidentified aircraft can be seen flying overhead. Jim Swale is fourth from right. (Photo J. Swale)

48. Commonwealth aircrew (wearing dark uniforms) and ground crew on a bright winter's day. F/Sgt Hugh Rice of Witham (middle row, extreme left) was a Flight Mech. Eng. with 295 Squadron. Ken Nolan front row, second left. (Photo H. Rice)

49. Stirling LK246 8Z-S flown by Wing Commander R. E. Angell DFC the CO of 295 Sqdn. The pennant below the nose signifies a commander's aircraft. The glazed panels above the bomb aimer's panel are unusual. (Photo via K. Fisher)

50. 'The Bushwacker' LJ995 8Z-H at snowbound dispersal. Aircraft still carries D-Day stripes and has sixteen missions painted on the nose. (Photo A. Murrie)

51. Rivenhall was never a naval station but four members of the senior service are in this picture! They are believed to have been drafted in to replace a shortage of electricians. (Photo A. Murrie)

52. Sergeant pilots of the Glider Pilot Regiment form the honour guard for the visit of American General Brereton on 9 April 1945. Behind the General is Major Mick Powell, the CO of the GPR at Rivenhall, and extreme left is Air Vice Marshal Scarlett-Streatfield, the AOC No. 38 Group. General Brereton awarded twenty-three decorations to aircrew who took part in Operation MARKET GARDEN in September 1944. (Photo M. Powell)

Above and below: 53. & 54. Horsa gliders on Landing Zone 'P' near Hamminkelm during Operation VARSITY on 24 March 1945. Troops can be seen through the dust and smoke which obscured a great deal of the battle area. (Photo via L. Archer)

Opposite above: 55. Happy smiles from members of 295 Sqdn 'A' flight as they cluster around 'Glorious Beer' LK129 8Z-B. Warrant Officer Peter Baldock of Braintree is standing sixth from right. The mission symbols are as follows: Dagger (supply drop), Mountain (supply drop in Norway), Windmill (supply drop in Holland), Bomb (close support), Glider (Operation VARSITY). (Photo P. Baldock)

Above: **56.** A total of forty-one Stirlings line up on the perimeter track for Operation DOOMSDAY on 11 May 1945. A further six Stirlings are at their dispersal points and twenty-four Horsa gliders can be counted on the original photograph. The village of Silver End is hidden by the morning mist and lies at the top of the picture. (Photo P. Baldock)

57. Junkers Ju188 D-2 photographed at the ex-Luftwaffe airfield of Gardermoen, Norway, in May 1945. Taken during a visit by the Rivenhall Stirlings which ferried supplies to the occupying troops. (Photo via P. Baldock)

58. Pierced Steel Planking (PSP) left behind when the squadrons departed, now used at Allshott's Farm to fence in the pigs. (Photo B. A. Stait)

Above: **59.** An air-raid shelter at Allshott's Farm, site of the Station Sick Quarters. (Photo B. A. Stait)

Right: **60.** This building was used to house the top-secret Norden bomb sight used by the 397th. Bomb Group. Note the misspelling of 'Bomb Site'. The photograph ooverleaf (61), shows the steel door inside the building.

Left: **61.** A total of eighty-four red and yellow bomb symbols to record the missions flown from Rivenhall by the 397th Bomb Group. (Photos B. A. Stait)

Below: **62.** The author (right) and Ken Fisher of Braintree with the remains of a V1 flying bomb, on the site of the motor pool building which was used by Carol de Coverley of Kelvedon as a welding shop until 1982. (Photo B. A. Stait)

63. This rusting shackle, used for dropping containers to the resistance groups by parachute, was found in one of the dried-up ponds on the edge of the airfield. (Photo B. A. Stait)

64. .50-calibre shells turned up by the winter ploughing from the area around the shooting butts. On the right is a Very Cartridge cap which was found near the site of the control tower. (Photo B. A. Stait)

65. Carol de Coverley, who organised the Rivenhall reunions, with a blackboard salvaged from the Station HQ showing the airfield layout. (Photo B. A. Stait)

66. John Ambrose at Woodhouse Farm in 1980. Note the wire-mesh panels which were used for aircraft parking on soft ground. (Photo B. A. Stait)

67. These wartime Nissen huts at the back of Sheepcote Farm are now used as a canteen for GEC Marconi workers. A 1975 photograph. (Photo B. A. Stait)

68. The cinema (originally the gymnasium) on the Polish Camp in 1975. (Photo B. A. Stait)

69. Type T2 hangar on the west side of the airfield, used for radar assembly and testing and much modified by GEC Marconi in the 1960s. (Photo GEC Marconi Ltd)

70. Rivenhall today. Sheepcote Farm and T2 hangar in the late evening sun during the summer of 1982. (Photo via C. de Coverley)

APPENDIX 1

Mission List of the 397th BG from 20 April to 4 August 1944

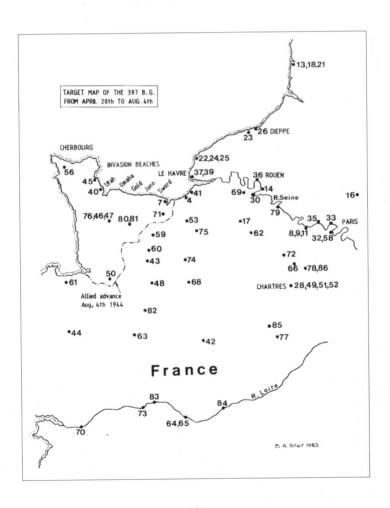

MISSION		DATE	TARGET	CODE	REMARKS	LEAD PILOTS
1	APRIL	20	LE PLOUY FERME – Pas de Calais	NB		COINER : DEMPSTER
2		21	BOIS DE COUPELLE – Pas de Calais	NB		COINER : DEMPSTER
3		22	VACQUERIETTE – Pas de Calais	NB		COINER : BERKENKAMP
4		23	BENERVILLE	CD		COINER : WINNINGHAM
5		25	BOIS COQUEREL – Somme	NB	* POOR VISIBILITY	COINER : DEMPSTER
6		26	ST. GHISLAIN – Near Mons	MY		DEMPSTER : WOOD
7		27	OUISTREHAM	CD		ALLEN : McLEOD
8		28	MANTES GASSICOURT	MY	* 10/10 WEATHER	McLEOD : BERKENKAMP
9		29	MANTES GASSICOURT	MY	* WEATHER (RECALLED)	WINNINGHAM : COINER
10		30	LOTTINGHEM – Pas de Calais	NB		ALLEN : WINNINGHAM
11	MAY	1	MANTES GASSICOURT	MY/B		BERKENKAMP : McCLEOD
12		2	BUSIGNY – Near Cambrai	MY		DEMPSTER : COINER
13		4	ETAPLES – Near Boulogne	CD		WOOD : GIBSON
14		8	OISSEL	RRB		BERKENKAMP : McLEOD
15		9	LE GRISMONT – Pas de Calais	NB		BRONSON : WINNINGHAM

16		10	CREIL	MY		WOOD ; BERKENKAMP
17		11	BEAUMONT LE ROGER	AF		DEMPSTER : BERKENKAMP
18		12	ETAPLES	CD		McLEOD : LOCKARD : WELTZIN
19		13	GRAVELINES – Near Dunkirk	CD		ALLEN : WOOD : RHODES
20		15	DENAINIPROUVY	AF	* 10/10 WEATHER	McLEOD : HAMILTON
21		19	ETAPLES	CD		McLEOD : LOCKARD : TAYLOR
22		20	ST. MARIE AU BOSC	CD		GARRETSON : RHODES
23		20	VARENGEVILLE SUR MER	CD		GARRETSON : McLEOD
24		22	ST. MARIE AU BOSC	CD	PFF (Pathfinder)	ALLEN : DEMPSTER
25		24	ST. MARIE AU BOSC	CD	PFF	McLEOD : DEMPSTER
26		24	DIEPPE (Harbour installation)			BERKENKAMP : TAYLOR: McLEOD
27		25	LIEGE	RRB		ALLEN : COINER
28		26	CHARTRES	AF		TAYLOR : WOOD
29		27	LE MANOIR	RRB		McLEOD : ALLEN
30		27	ORIVAL	RRB		BERKENKAMP : McLEOD
31		28	LIEGE	RRB		McLEOD : DEMPSTER
32		28	MAISONS LAFFITTE	RRB		GARRETSON

33		29	CONFLANS	RRB		McLEOD : BRONSON
34		29	BEAUVOIR	NB		BERKENKAMP : LOCKARD
35		30	MEULAN	B		DEMPSTER : PEMBERTON
36		31	ROUEN	B		McLEOD : BRONSON
37	JUNE	1	LE HAVRE	CD		DEMPSTER : McLEOD
38		2	CAMIERS	CD		ALLEN: RHODES : STEPHENSEN
39		3	LE HAVRE	CD		BERKENKAMP : TAYLOR
40		6	DUNES DE VARREVILLE (Utah)	CD	D-DAY	McLEOD : BERKENKAMP : ALLEN
41		6	TROUVILLE	CD	D-DAY	DEMPSTER : WELTZIN
42		7	LE MANS	RRB		BERKENKAMP : ALLEN
43		7	FLERS	MY		DEMPSTER : McLEOD
44		8	RENNES	RRB	Bombed casual targets	ALLEN : BERKENKAMP
45		10	QUINEVILLE	CD		BERKENKAMP : ALLEN
46		11	ST. LO (Highway)		* 10/10 WEATHER	BERKENKAMP : WOOD
47		12	ST. LO (Road junction)	B		ALLEN : DEMPSTER
48		13	FORET D'ANDAINE	SD		McLEOD : BERKENKAMP
49		14	CHARTRES	RRB		WOOD : ALLEN

50		14	ST. HILAIRE	B		DEMPSTER : McLEOD
51		15	CHARTRES	B		BERKENKAMP : WOOD
52		17	CHARTRES	B		DEMPSTER : ALLEN
53		18	MEZIDON	MY	* 10/10 WEATHER	BERKENKAMP : McLEOD
54		18	BACHIMONT – Pas de Calais	NB	PFF	BERKENKAMP : McLEOD
55		20	GORENFLOS – Somme	NB	* 10/10 WEATHER	WELTZIN : BUCKLER
56		22	NOUAINVILLE	DA	PFF	McLEOD : LOCKARD
57		23	LAMBUS – Pas de Calais	NB	PFF	ALLEN : LOCKARD
58		24	MAISONS LAFFITTE	RRB		BERKENKAMP : STEPHENSEN
59		30	THURY HARCOURT	B	* 10/10 WEATHER	DEMPSTER : BRONSON
60		30	CONDE SUR NOIREAU	RJ	PFF	McLEOD : BERGER
61	JULY	6	DOL TO RENNES (Railroad)			COINER : LOCKARD
62		6	FORET DE CONCHES	FD		McLEOD : WELTZIN
63		7	LAVAL AREA – USSY (Motor transport)		* WEATHER	WOOD : BRONSON
64		8	SAUMUR	RRB		BERKENKAMP : BERGER
65		8	SAUMUR	RRB	* 10/10 WEATHER	ALLEN : RHODES
66		9	NOGENT LE ROI	RRB		WOOD : HAMILTON

67		11	CHATEAU DE TERTU	FD		BERKENKAMP : STEPHENSEN
68		12	FORET D'ECOUVES	FD	PFF	DEMPSTER : WELTZIN
69		16	BOISSEY LA LONDE	RRB	PFF	ALLEN : HAMILTON
70		16	NANTES	RRB		WOOD : GARRETSON
71		18	DEMOUVILLE – Near Caen		OPERATION GOODWOOD	DEMPSTER : ALLEN
72		18	CHERISY	RRB		BERKENKAMP : STEPHENSEN
73		19	LA POSSONNIERE	RRB		ALLEN : BERGER
74		23	ARGENTAN	RRB	PFF	DEMPSTER : HAMILTON
75		24	LIVAROT	FD	PFF	McLEOD : STEPHENSEN
76		25	MONTREUIL – Area bombing		OPERATION COBRA	ALLEN : WELTZIN
77		25	CLOYES	RRB		McLEOD : GARRETSON
78		26	EPERNON	RRB		ALLEN : BERGER : WEST
79		28	COURCELLES	RRB	PFF	BERKENKAMP : WELTZIN
80		30	CAUMONT	DA	PFF	McLEOD : HAMILTON
81		30	CAUMONT	DA	PFF	WOOD : GARRETSON
82		31	MAYENNE (Viaduct)	RRB	PFF	COINER : WELTZIN
83	AUGUST	1	LES PONTS DE CE	RRB		McLEOD : WEST : TAYLOR
84		2	CINQMARS	RRB		WELTZIN : LOCKARD

85		3	COURTALAIN	RRB		WOOD : GARRETSON : ROBERTS
86		4	EPERNON	RRB		DEMPSTER : RHODES

KEY

NB	Noball
CD	Coast defences
MY	Marshalling yard
AF	Airfield
B	Bridge
RRB	Rail road bridge
AD	Ammunition dump
FD	Fuel dump
SD	Supply dump
*	No bombing

Serial Numbers
of 397th BG Marauders

Serial 42–96	Sqdn. Code	Aircraft letter	PILOT	AIRCRAFT NAME	REMARKS
052	9F	J	BRADEN		
78					CRASHED 17.6.44
083	6B	–	GROSS		SHOT DOWN 9.8.44
086					CRASHED 11.3.45
088	U2	P	PATTERSON	OLD GRUESOME	
091	6B	B	McCARTHY		
093	X2	V			CRASHED 28.7.44
110	6B		KRETSCHNER		CRASHED 7.6.44
111	9F	U	LEAVERTON		
114	9F	D	FLOWERS	DINA	
115	9F	Q	SCHWARZROCK	THE REAL McCOY	SHOT DOWN 15.2.45
116					SHOT DOWN 4.8.44

118	9F	T	BOOR	OLD BOOMERANG	SHOT DOWN 7.8.44
120	9F	R		MAMA LIZ	SHOT DOWN 24.6.44
121	9F	–	KNOX		SHOT DOWN 24.6.44
122	9F	P	HURLEY		SHOT DOWN 13.3.45
123	9F	O	CORDELL	SPARE PARTS	CRASHED
124	9F	M	GATEWOOD	HOLY MOSES	CRASHED 26.2.45
125	9F	L	GLEIS		
127	9F	–	NEILL		SHOT DOWN 24.6.44
129	U2	A	ALLEN	SEAWOLF II	SHOT DOWN 7.7.44
130	U2	R	HOWARD		
132	U2	B	COINER	LUCKY STAR	CRASHED 17.6.44
133					SHOT DOWN 24.6.44 IN FRIENDLY TERRITORY
134	6B	N	DEMPSTER	COLLECT ON DELIVERY	SHOT DOWN 25.12.44
135					CRASHED 17.6.44
136	X2	Z			SHOT DOWN 25.12.44
137	9F	Y	WOOD		SHOT DOWN 13.5.44
138	U2	C	WEST	BY GOLLY	SHOT DOWN 16.7.44
139	U2	D	SMITH	SUSAN J	SHOT DOWN 23.12.44
140	U2	F	BERKENKAMP		
141	U2	K	BERGMAN		
142	X2	A	McLEOD	DEE FEATER	CRASHED 10.8.44
143	X2	–	FREEMAN		SHOT DOWN 8.5.44
144	X2	C	ESTES	BANK NITE BETTY	SHOT DOWN 23.12.44

145	U2	F			
146	X2	F		MAMMY YOKUM II	
147	X2	G	ILLANES		
148	X2	H	COLAHAN		SHOT DOWN 8.4.45
149	X2	J	HOCH		
150	X2	K	HAYES		
151	X2	L	BROWN	TACONITE EXPRESS	
152	X2	M	PARKER	MISSOURI MULE II	
154	X2	O	ROBERTS		
155	X2	P	BARCROFT		
156	X2	W	JORDAN		
157	U2	H	RYHERD	SHARON ROZANNE II	
158	U2	J	RICHARDSON	LITTLE PEEDOFF	
159					SHOT DOWN 23.12.44
160	U2	L	NORTH	BILLIE WILLIE V	SHOT DOWN 18.3.45
161	U2	M	STANGLE	PATTY KAY	SHOT DOWN 24.6.44 FRIENDLY
162	U2	N	QUIGGLE	DOTTIE DEE	
163	6B	O	SPAULDING	MAMA LIZ II	
165	6B	T	SHAEFFER	BIG HAIRY BIRD	
166	6B	S	GROSS		CRASHED 24.5.44 UK
167	6B	X			
168	6B	P	GARRETSON		COLLISION 3.8.44

170	6B	G	CRABTREE		
171	U2	S	SENART	BAR FLY	
172	U2	R	COOK OR TURNER		CRASHED 13.1.45
174	6B	Q	CRAVEN		CRASHED 11.5.44 UK
175	6B	V			?CRASHED 8.5.44 UK
177	6B	–	POWERS		SHOT DOWN 24.6.44
178	6B	M	WILLEMSEN		
181	6B	A	STEPHENSEN	DRAGGIN' LADY	
182	6B	K			SHOT DOWN 23.12.44
183	6B	S	WILLEMSEN		CRASHED 11.1.45
185					SHOT DOWN 23.12.44
186	9F	Y	PEMBERTON	FLAME McGOON	
188	6B	–	THOMPSON	LASSIE COME HOME	SHOT DOWN 10.8.44
191	9F	N	OVERBEY	THE MILK RUN SPECIAL	
193	6B	–	KRETSCHMER		CRASHED 9.7.44 UK
197	X2	–	POLLACK		CRASHED 10.3.45
201	6B	L		BLIND DATE	SHOT DOWN 23.12.44
204	X2	X	COLAHAN		CRASHED 25.12.44
205	X2	S	BOYAR		
219	9F	H	STONER		CRASHED 1.12.44
221					SHOT DOWN 23.12.44
222	U2	R	McEACHERN		
253	6B	B	McCARTHY		

278					SHOT DOWN 11.8.44
280	9F	T	WHITMIRE	BABY BUTCH II	SHOT DOWN 23.12.44
283	6B	Y	McCORKLE		
288	X2	Q	HOCH		CRASHED 30.11.44
289					SHOT DOWN 9.8.44
290					CRASHED 2.12.44
43–34	Possibly 253–288				
117	U2	S	STEERE	HI HO SILVER	
118	U2	M	PATTERSON		
122	U2	G	SILVERBACH		
124	U2	B	HELLSTROM		
125	U2	T	TAYLOR		
126	U2	H	WEST		SHOT DOWN 1.8.44
127	U2	D	BITZER		
128	U2	P	THORNTON		
130	U2	O	HOWARD		
139	6B	Z			
172	U2	H			
188	6B	–	CRABTREE		SHOT DOWN 10.8.44
217	U2	Q			
288	X2	J	KING	JUST DREAMIN'	
312					CRASHED 11.8.44
401	9F	K	WOOD	HELEN HIWATER II	
430					SHOT DOWN 23.12.44

434					SHOT DOWN 23.12.44
450					CRASHED 19.4.45
566	9F	O		EL LOBO	
984					SHOT DOWN 13.3.45

X2 – 596 Sqdn; 9F – 597 Sqdn; U2 – 598 Sqdn; 6B – 599 Sqdn.

This list of 397th BG Marauders has been compiled with the help of Roger Freeman and the late Jim Snow plus a great deal of detective work in the official records. Unfortunately there appear to be several contradictions in the records and they cannot be entirely relied upon. The author apologises for any inaccuracies.

Pilots were generally associated with one particular aircraft, which became 'their own' plane and were responsible for naming it where this was done. However, depending on circumstances, pilots and crews might be called on to fly 'strange' aircraft. Regular crews were particularly unhappy if 'their' aircraft was lost or damaged when flown by another crew. Pilots' ranks have not been given as their status was altered by rapid promotions while stationed at Rivenhall.

Map of Landing Zones
on 24 March 1945

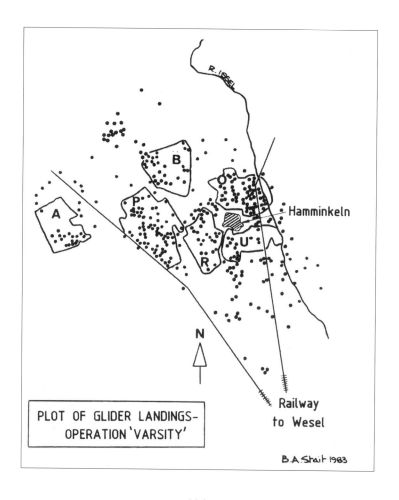

PLOT OF GLIDER LANDINGS—
OPERATION 'VARSITY'

Stirlings of 295 Squadron which Took Part in Operation VARSITY – 24 March 1945

Aircraft Serial	Squadron Code	Take-off	Landing	A/C Captain
LK246	8Z-S	07.00	12.00	W/Cdr R. E. Angell
LJ652	8E-X		12.05	F/Lt R. Churchill
LJ576	8Z-E or D		12.10	F/O T. Sessions
LK129	8Z-B		12.10	F/Lt A. M. Scott
LK355	8E-Y		12.10	F/O K. Newton
LK558	8Z-D or 8Z-F		12.15	S/Ldr Stewart
LK122	8Z-D or 8Z-F		12.15	F/Sgt S. A. Currah
LJ922	8Z-F or I		12.20	F/Sgt A Knott
LK137	8E-J			W/O Symmons (Shot down 10.33)
LK346	8Z-H	07.05	12.05	S/Ldr H. F. Johnson
LK136	8E-V		12.15	F/Lt E. Pearson
LK495			12.25	P/O H. Stuchbury
LK513	82-Q	07.10	12.10	F/O Finlay
LK120	8Z-W		12.10	F/O T. Taylor

LK287	8Z-C			F/O A. G. Smith (Returned early)
LK290	8Z-Z or 8E-Z		12.10	F/Lt F. Jones
LJ950	8Z-U		12.15	F/O R. C. Skipton
LK553 or LK330	8Z-P		12.20	F/Lt G. A. Watson
LK543	8Z-K		12.30	P/O L. Bellinger
LJ929	8E-T		12.30	F/O P. G. Hillyer (NZ)
PK226	8Z-R		12.30	W/O D. Draper
LK575	8Z-A		12.35	F/Lt F. Starling
LJ890	8E-L		08.15	F/O Webster (Returned early)
LK144	8E-M	07.15	12.15	F/Sgt Storer
LK132	8Z-N		12.20	F/O R. A. Sloan
LJ976	8E-Q		12.35	F/Sgt C. Hall
EF446	8Z-O or 8E-O		12.40	F/Sgt Dyson
LJ951	8E-C	07.20	12.20	W/O E. Dax
LJ591?			12.25	F/Lt R. Scott
LK288	8Z-G		12.25	F/O R. W. Bowman
LK351	8E-N	08.55	13.05	F/Sgt Sabourin

Although the take-off and landing block times are 'as recorded' they are based on the time-honoured 'nearest five minutes' system so are not therefore precisely accurate. The aircraft in each time block are not necessarily in the order of take-off.

This deceptively simple list is a graphic example of the difficulties experienced when researching from official records.

295 Sqdn Operations Record Book held in the Public Record Office is the original source but unfortunately it contains a number of errors; eleven of the serial numbers are either incorrect or

incomplete and there are two pairs of identical serial numbers. The squadron codes have been added by the author from aircrew log books which are themselves not always free from errors. My thanks to Bryce Gomersall, who compiled *The Stirling File*, for assistance with this appendix.

Stirlings of 570 Squadron which took part in Operation VARSITY, 24 March 1945:

Aircraft Serial	Squadron Code	Take-off	Landing	A/C Captain
PW406	E7-I	07.20	12.30	P/O Young
LK202 (1)	E7-X	07.20	12.30	W/O Pritchard
LK549	E7-J	07.24	12.29	W/O Kirkham
LJ620 (6)	E7-O	07.25	12.12	F/O Momrun (2)
LJ640	V8-B	07.37	12.37	W/CDR Bangay
LK156	V8-I	07.37	12.47	F/O Shuter
PK234	V8-E	07.40	13.00	F/Lt Spafford RCAF
LJ622	V8-M	07.40	12.45	F/O Merritt
LK190	V8-J	07.40	12.40	F/O Houlgate
LK291	V8-A	07.40	12.50	W/O Narramore
LJ596 (3)	V8-K	07.40	12.45	W/O Bentley
LK199	V8-L	07.45	13.10	F/Lt Brierly
LJ616	V8-D	07.45	12.45	F/O Rogers RAAF
LK559	V8-Q	07.45	13.00	F/O Murphy
LJ615	V8-P	07.45	12.50	F/O Downing
LJ992	V8-N	07.45	13.00	F/O Parkinson
LK280	V8-F	07.47	12.52	F/O Campbell
LK289	V8-G	07.50	13.10	F/O Hicks RAAF

LJ650	E7-W	07.50	13.00	P/O L. Drife
LK555	E7-S	07.53	13.08	S/Ldr Stewart
LK286 (4)	E7-T	07.55	12.45	F/O McDonald RCAF
LJ612	E7-L	07.55	13.00	F/O Davison RCAF
LJ645 (5)	E7-M	07.56	12.45	F/Lt Burkby
LK292	E7-V	08.01	12.46	F/O Jennings RNZAF
LJ620 (6)	E7-O	08.02	12.40	P/O Hincks
LK154	E7-P	08.05	12.45	F/Lt Bullen
PW255	E7-K	08.06	12.54	W/O Marshall
LJ636	E7-N	08.15	12.35	F/O Burr
	Y8-Y	08.05	13.05	F/Lt Sharp

Extract from 570 Squadron Operations Record Book in Public Record Office:

(1)   U/C collapsed on take-off 18.4.45
(2)   Pilot killed 22.4.45 in LJ645
(3)   A/C lost on 20.4.45
(4)   Crashed on landing 2.4.45
(5)   Shot down over Aarhus 22/23.4.45
(6)   Discrepancy

Summary of Loads Carried by 295 and 570 Sqdns for Operation VARSITY – 24 March 1945

L.Z.	SQDN	A/C	LOAD					Broke tow rope	Flak damage	REMARKS
			Troops	Jeeps	Trailers	M/Cycles	Hand carts			
P	295	5	94	2	3	–	2	–	–	
P	570	30	257	22	23	5	2	2	3	
R	295	14	29	6	3	–	5	1	2	
U	295	11	103	10	7	–	–	–	2	One shot down 5 crew returned

Equivalent Commissioned Ranks

US Army Air Force	Royal Air Force
General of the AF	Marshal of the RAF
General	Air Chief Marshal
Lt General	Air Marshal
Major General	Air Vice Marshal
Brig. General	Air Commodore
Colonel	Group Captain
Lt Colonel	Wing Commander
Major	Squadron Leader
Captain	Flight Lieutenant
Lieutenant	Flying Officer
Second Lt	Pilot Officer

Postscript

The 397th Bomb Group Association maintain a thriving organisation which caters for their ex-comrades. In recent times they seem to have gone from strength to strength and several reunions have been held in different parts of the United States during the past few years.

While this book was in its proof stage the Association sent the author details of a trip to Europe planned for 1984. The veterans and their families intended to visit many of their former bases on the Continent concluding their tour at Rivenhall on 17 June.

The author is not aware of any reunions held by the 363rd Mustang Group nor by 295 and 570 RAF squadrons. However, in 1979 Mr Carol de Coverley (a Kelvedon man who runs a small welding business on the airfield) organised two reunions at Rivenhall which caused a great deal of local interest. A number of ex-295 and 570 personnel turned up and it is hoped to repeat the success in 1984.

Bibliography

Air Britain, *The Stirling File*.

Allen, Peter, *One More River*.

Beck, Henry C. (ed.) *The 397th Bomb Group (M) Bridge Busters* (1946).

Bowyer, Mike, *The Stirling Bomber*.

Collier, Basil, *Battle of the V Weapons*.

Freeman, Roger, *Mustang at War*.

Freeman, Roger, *Marauder at War*.

Freeman, Roger, *Camouflage and Markings* No. 14. (Marauder markings).

Lloyd, Alan, *The Gliders*.

Rust, Kenn, *Ninth Air Force in World War II*.

The Martin Marauder and the Franklin Allens. Sunflower University Press (Letters from Lt. Col. Allen (CO of the 598th Bomb Sqdn.) to his wife).

Wright, Lawrence, *The Wooden Sword* (Glider ops. inc. 295/570 refs.)

AeroAlbum – Summer 1968. Kenn Rust (article 397 BG).

Aeronautics, September and October 1944 (articles 397 BG).

The Public Record Office:

   AIR 27/1644, AIR 27/2041 and 2042 (295 and 570 sqdn. ops. records).

   AIR 28/666 (Rivenhall station operation record book).

Maxwell Air Force Base, Alabama, USA.

The Archives Branch of the Albert F. Simpson Historical Research Center for the supply of microfilms containing the history of the four squadrons of the 397 BG and the three squadrons. of the 363 FG.

The following magazine articles by Jack Stovall of Memphis relating to the 397th BG:

   Air Progress Aviation Review – Summer 1979 (Saga of 'By Golly')

   Air Classics – June 1979 (Marauder Men).

   Air Classics – October 1979 (When Marauders Flew for Hollywood – the Making of the Film *A Guy Named Joe*).

   Air Classics – April and May 1980 (Diary of Neil McGinnis, 598 BS gunner).

The following magazine articles by editor/publisher Steve Blake:

   Fighter Pilots in Aerial Combat – numbers 5, 6 and 7 (the 363rd Fighter Group in World War II).

Photo Credits

Many individuals answered my plea for photographs taken at Rivenhall and I express my grateful thanks to them all. Each photo has been credited to the person who supplied it, although in some cases they originated from an unknown source. Most of the 397th BG photos were from the Photo Officer, Jim Snow, who lent the original 5 inch x 4 inch negatives to the author in 1977, just before his death the following year. Steve Blake, the editor and publisher of *Fighter Pilots in Aerial Combat*, kindly lent me his collection of 363rd FG negatives which had come from several 363rd men and gave permission to quote from articles in the magazine.

Acknowledgements

A great many people have helped in the compilation of this history and I am indebted to them all. I am particularly grateful to Jack Stovall of Memphis USA who supplied many contacts and a wealth of information from both the 397th BG and the 363rd FG and with whom I have developed a lasting and deep friendship. In addition, Ken Fisher and Carol de Coverley (both of whom live not far from Rivenhall) have kept up a non-stop exchange of ideas and inspiration over many years; many thanks to them both. I must record my gratitude to the GEC/Marconi company for permission to photograph the airfield and its environs and to interview the employees and my special thanks to the company historian Mrs B. Hance for providing much useful information. The following American servicemen assisted with photos and information: Bud Udick, Jim Snow, George Parker, Moses Gatewood, Jim Russell, Manley Nelson; sadly not all those with whom I have corresponded are still with us.

On the British side I wish to thank especially Peter Baldock, Jim Swale, Tony Wadley, E. A. Woodger, A. Lowe, Mrs I. Doughton, John Spurgeon, and a special thank you to Leonard Archer (who flew on every one of the three airborne ops. – Normandy, Arnhem, Rhine crossing) for lending me his copies of VARSITY reports. Many other friends and associates have helped in various ways; Roger Freeman (whose book *The Mighty Eighth* really sparked off my interest in researching Rivenhall airfield), Ken Draper, John Hunt, Arthur Lane, 'Jimmy' James, Ms P. L. Hargreaves, Ms H. M. Daciow, Mrs L. Palmer, Mrs H. Smith and Mrs L. A. Whitaker. Too numerous to mention are the many people who replied to my requests for help in the local paper and the RAFA magazine *Airmail* but I am grateful to them all. Even if their individual stories were excluded from the finished MS they still helped me to get a picture of life at Rivenhall from 1944 to 1946.

Finally I wish to extend my grateful thanks to a devoted aviation buff, Jack Meaden, who read the final draft and offered many useful suggestions and helped to set the record straight. Any remaining inaccuracies are the responsibility of the author.